ACQUIRED TASTES

Food, Health, and the Environment

Series Editor: Robert Gottlieb, Henry R. Luce Professor of Urban and Environmental Policy, Occidental College

ACQUIRED TASTES

STORIES ABOUT THE ORIGINS OF MODERN FOOD

EDITED BY BENJAMIN R. COHEN,
MICHAEL S. KIDECKEL, AND ANNA ZEIDE

THE MIT PRESS   CAMBRIDGE, MASSACHUSETTS   LONDON, ENGLAND

This book was set in Avenir by Westchester Publishing Services, Danbury, CT. Printed and bound in the United States of America.

Library of Congress Cataloging-in-Publication Data

Names: Cohen, Benjamin R., editor. | Kideckel, Michael S., editor. | Zeide, Anna, 1984– editor.
Title: Acquired tastes : stories about the origins of modern food / edited by Benjamin R. Cohen, Michael S. Kideckel, and Anna Zeide.
Description: Cambridge, Massachusetts : The MIT Press, [2021] | Series: Food, health, and the environment | Includes bibliographical references and index.
Identifiers: LCCN 2020049512 | ISBN 9780262542913 (paperback)
Subjects: LCSH: Food—History. | Food habits—History.
Classification: LCC GT2860 .A29 2021 | DDC 641.3009—dc23
LC record available at https://lccn.loc.gov/2020049512

10 9 8 7 6 5 4 3 2 1

CONTENTS

SERIES FOREWORD

Acquired Tastes: Stories about the Origins of Modern Food is the nineteenth book in the Food, Health, and the Environment series. The series explores the global and local dimensions of food systems and the issues of access, social, environmental and food justice, and community well-being. Books in the series focus on how and where food is grown, manufactured, distributed, sold, and consumed. They address questions of power and control, social movements and organizing strategies, and the health, environmental, social and economic factors embedded in food-system choices and outcomes. As this book demonstrates, the focus is not only on food security and well-being but also on economic, political, and cultural factors and regional, state, national, and international policy decisions. Food, Health, and the Environment books therefore provide a window into the public debates, alternative and existing discourses, and multidisciplinary perspectives that have made food systems and their connections to health and the environment critically important subjects of study and for social and policy change.

Robert Gottlieb, Occidental College
Series Editor (gottlieb@oxy.edu)

INTRODUCTION

An eater sits down to a morning meal nearly a century ago. She spreads her slice of processed white bread toast with a thick layer of strawberry jam, sweetened with corn syrup. Her bowl of Shredded Wheat comes from a box emblazoned with the natural majesty of Niagara Falls. An imported tropical banana lies alongside the bowl. She idly stirs a tablespoon of bright white sugar into her morning cup of coffee as she reads the newspaper, musing about her busy day ahead.

At the supper table that evening, as a special treat, she opens a Pepsi-Cola for herself, enjoying its fizzy flavor and the soothing effect on her stomach. Her dinner companion opens a glass bottle of beer—a light, refreshing pilsner. They set the table and, when it's time to eat, she brings out a pot roast. As she walks into the dining room, she recounts her experience at the store that day. The grocer had shown her a can of "vegetable meat," called Protose. "Imagine that!," she exclaims. Bowls of canned green beans and corn dressed in a cream sauce accompany the roast, dishes she made following a recipe from a favorite cookbook author promising updated versions of grandma's recipes. And for dessert, cold, crisp watermelon from the icebox, not yet mushy.

This cook has planned this meal according to her understanding of nutrition, balancing the prepared dishes in line with domestic science. She sees her food and presentation as part of who she is, and food's abundance and safety as evidence of a modern way of life in the first half of the twentieth century.

Today, these imaginary prewar meals and some of the brands seem familiar to us. And yet, processed, packaged, labeled, and advertised foods would have been unimaginable just a half century before the meal described above. In the decades before World War II they became available throughout the United States and Europe. Radically new ingredients and foods emerged from a set of social phenomena that defined the turn of the twentieth century—a combination of globalized trade, colonial efforts, and imperial politics.

Our imaginary cook and eater was just one kind of consumer in the early twentieth century. It might be tempting to consider her typical, especially as the era was rife with efforts to define and manage the "average" citizen, but that could be a contrivance too far. The kind of food we eat today varies by region, social class, and ethnic and racial identities, among other social categories; so too did it a century ago. But this idea of a meal is an invitation to consider the power differences and conflicts that coalesced in food in the early twentieth century, bearing the weight of the prior fifty years. That foundation in the construction of modern food is still with us today.

Although popular writing of the twenty-first century may lead us to think of the postwar period as the beginning of some idea of a "modern" food system, we need to look earlier. By the 1930s, the chief features that most people consider part of today's food system had already taken form, shaping what came to rest on our imagined table. The wide variety of foods, the nutritional and scientific considerations, the connection to identity, the whiteness of the bread and sugar, the availability of fresh produce like bananas and watermelon, and the industrialized nature of so many items—the corn-syrupy jam, cereal, beer, Pepsi, fake meat, and canned food—all illustrate a significant shift from earlier ways of eating toward a food system that most readers recognize today. We'll learn more about each of these foods and features in the essays that follow.

Our foods and farms are products not just of factories and fields, but of history. The options on our tables and store shelves today carry the weight of the stories they gathered along the way. For the sake of a healthier and more sustainable future, one with greater attention to justice, equity, nutrition, and ecological integrity, we will need policies and environmental actions that consider those stories and understand how they can be remade. We need to base future policies on a clearer picture of the assumptions and values built into our food system in the first place.

To say that food is the product of history may be stating the obvious, but it bears repeating. In this collection, we want to emphasize two main points—both rooted in concerns long central to the discipline of history: the importance of historical particularities, and the centrality of storytelling. First, we seek to move away from World War II as the pivot point for the establishment of modern food in the United States and Europe and instead show the crucial role of the half century between the 1870s and the 1930s. Second, we do so with careful attention to how we tell our stories here, and how the broader story of modern food is often told.[1] If, as Aristotle argued in the *Poetics*, all stories have a beginning, a middle, and an end, our point is that the World War II era is the middle, not the beginning— and that the end is not yet in sight.[2]

Many of us sense that food as we know it wasn't always this way. Stop for a casual conversation at a farmers' market or gather with groups intent on reforming the food system, and you'll hear about some hazy past when we ate differently than we do today. In those conversations, we perhaps sense that the complex food supply chains, enormous grocery stores, invisibility of the labor making the food, and availability of diverse ingredients from around the world are all somewhat new—that they had to come from *somewhere*. But where and when is less clear.

When food writers and reformers do specify a time and place, they often pin the start of modern food to the postwar era when farms grew bigger, chemicals from munitions plants got redirected to pesticide and fertilizer production, and government policies further encouraged consolidation and industrialization. A typical bookshelf on American and European food might offer a common template of familiar consumer examples. While one popular book cites Cheez Whiz, Frosted Flakes, and Froot Loops, the next highlights Chef Boyardee, Rice-a-Roni, and Chex as another points to Lunchables, Shake 'n Bake, and Oreos. References to TV dinners, microwavable food, prepacked meals, and sugary breakfast cereal with cartoon mascots and action figures direct readers to where it all went wrong. Speaking of sugar, one author tells readers it "all changed in the late 1960s." Another observer writes that "In the universe of processed food, World War II was the Big Bang." The postwar period is a common and available reference.[3] And certainly, there was much turbulence and change in that period.[4]

Perhaps no food writer has been more influential in this framing than Michael Pollan. In a 2006 essay he warned readers to stay away from anything that their "great-great-great grandmother wouldn't recognize as food." Two years later, the decree had softened and shortened, as Pollan's *In Defense of Food* (2008) asked worried consumers to go only as far back as their "great-grandmothers" for guidance. Two generations of change fell away with one editorial stroke, leaving us again with World War II as the defining turning point, with the suggestion that a few generations don't make much difference in pointing to historical origins.[5] While Pollan's deft prose and ability to bring new eaters to the cause have rightly been celebrated for some time, his historical assessments limit the ability of policymakers, historians, and consumers to think to the future without replicating the assumptions of the past. They provide a pleasingly simple story line that leaves readers with little stomach for the complicated realities of policymaking and reform.

But Pollan is not alone in this assessment and, in any case, has become a too-common target for scholars complaining about partial history or a wobbly moral compass. The skewed timing and emphases on the second half of the twentieth century likely also arose because the postwar period ushered in a host of important works about the problems of so-called modern food. Frances Moore Lappé's 1971 *Diet for a Small Planet*, to take one important example, led millions to question overzealous carnivorous habits, with the understanding that public and environmental health were both at risk. Alice Waters's famous culinary turn the same year put organic ingredients and connections to local farms in the spotlight, likewise elevating the public conversation around nutrition and kitchen practice. Wendell Berry's writing in the 1970s and after on agrarian identity and the damages of industrial farming for American land and culture added to a robust postwar discussion about the failings of modern food and agriculture. In each case, these voices were reacting to a food system that had been coming into place for decades.[6]

In truth, the World War II era was more of an intensification than a starting point, in a wide variety of arenas. Environmental historians have described that same time period as the font of a "great acceleration" in climate change that marks the historical period from roughly around the 1940s to today; a wide array of researchers would pin the "golden spike" of the Anthropocene to the creation of the atomic bomb. What such observers note is a kind of acceleration of features that had been put in place before World War II. We see this great acceleration in play with food systems too. By the mid-twentieth century, food in much of the United States and Europe bore the collective imprint of global trade, colonization, mass distribution, racist science, celebrity culture, and factory processing. In the mid-nineteenth century, it did not. The stories in this collection open a window onto how these processes changed the culinary and gustatory landscape.[7]

It thus isn't the place of Pollan's great-grandmother that's interesting; it's the editorial erasure of fifty years. We need to put those years back in place. This book brings together historians who have shown that our current food system began to come together during a period of immense colonial, imperial, scientific, and global development around the turn of the twentieth century. Echoing food historian and food studies pioneer Warren Belasco, the stories here offer a "sobering corrective to much of the nostalgia found in popular discourse, especially the common assertion that 'industrial agriculture' is largely a post–World War II invention." It was before the war that "uniformity, sterility, predictability—the values inherent in machine-age cuisine," as food historian Laura Shapiro put it, took root, making possible the growth of modern agricultural and food policy.[8]

We titled this collection with careful attention to the process and timing of the food system's formation. As the phrase is typically used, "acquired tastes" refers to foods that people come to like over time, such as bitter coffee or alcohol. Although most people dislike these foods on their first exposure, they often come to love—or perhaps depend on—them. In the same way, many of the changes that we call "modern" were resisted before becoming desires or crutches. We'd like to disrupt modern food's air of inevitability, making clear that the tastes many consider "normal" resulted from repeated processes conducted over time.

The dating of modern food's origins matters because imagining more effective policies requires better understanding the depth of commitment in existing ones. Communities that want to make change must reckon with their pasts as necessary foundations from which to move forward. As part of this, rethinking a food system especially prominent in the United States requires wrestling with the decisions people in the past made when building structures of food production, distribution, and consumption.

Trying to define "modern" food in singular or universal terms would be too restrictive. There are too many cultural practices, too many diets, and too much agricultural variability to claim a uniform system. We use the imperfect but helpful term "modern" to describe the size, bureaucracy, and anonymity of an ever-larger number of people's lives at the turn of the twentieth century. We think of "modern" as referring to the food system that exists now but did not in the past, and does not necessarily have to exist in the future. Changing our path into the future requires that we understand that our food and farms are our history because the past development of current modes of growing, cooking, and eating live with us still. It requires that we tell better stories.

The existing stories often spring from a proposed moral dichotomy: that our tables and shelves are bad today, so they must have been good before. Or they depend on a historical impulse that asks us to reverse time and go backward a few generations. They are almost always about historical decline, lamenting and lambasting a fall from grace and an expulsion from Eden.[9] Too often, curious advocates take that dictum to peer backward from the safety of the inevitable present, as if taking up the famous and almost always misinterpreted George Santayana dictum that those who cannot remember the past are condemned to repeat it.

We are after something different. Instead of looking back from the present, we propose taking the vantage point of the past and looking forward. Imagine sitting in an office chair that swivels. We could just turn the chair 180 degrees to look behind us. We want instead to pick up that chair and move it back to the later 1800s so we can look forward. Doing so helps us recognize that the future was not preordained, that decisions of the time were based on values and propositions then ascendant and now prominent. Doing so also means that in our efforts to narrate a selection of key elements of that history, we try to avoid moral dichotomies and either/or framing.

But we don't have a time machine. We can't simply move into the nineteenth century to understand past decisions around food. What we can do, though, is narrate the lives, options, circumstances, and arguments of people at that time. That approach shows how complicated the origins of modern food were. Food was both cause and effect of broader movements to create and meet demands in a growing and diverse public—to colonize and control distant lands, to build new markets, to secure global trade, and to define new ideas about nutrition, health, and sustenance. As the chapters to come show, those factors led the relationships that once governed people, places, and food to morph into celebrity endorsement, major branding, and scientific purity tests. And today's food system took form in certain ways because it emerged alongside the modern professional state, especially in the United States, at a moment surrounding 1900 when those with power trusted in expertise, institutions, and large organizations. If the current food system had started in the 1950s, it would look different, in part because, after World War II, a widely shared liberal faith in large institutions began to decline, not ramp up. In other words, the particular historical context matters.

The tool we employ to shift the World War II–centric narrative is the aggregation of smaller stories. These are stories both in the sense of an account of the past and in the spirit of attentiveness to narrative and plot. We have produced them to address three broad areas of modernization, largely in the United States, but also in the Caribbean, Europe, and the Philippines. Part I, "Time and Space," presents a set of five chapters that attend to distance, mobility, separation, and new connections. Part II, "Trust," addresses the questions of familiarity, suspicion, belief, and confidence that were newly in flux in great part due to the confusion exacerbated by patterns discussed in part I. Part III, "Science," describes the changes wrought by a growing commitment to institutional expertise that put food into racial, gendered, and related nutritional categories.

Our chapters tell of developments and choices that were neither inevitable nor uncontested. The transition to packaged food, uniformity, and predictability that so many people now live with relied on how millions of individuals felt they—and others—should eat.

Part I begins with a set of transnational histories that show how the collapse of distance proved vital to fashioning the modern food system. We begin with two tales of wheat, one of them about growing and shipping the staple commodity, the other about processing and baking it. In chapter 1, Tom Finger asks how California's Central Valley became one of the most productive agricultural zones in the world. He describes how the "agricultural lands" were made through the violent displacement of native peoples, the dubious land claims of prospectors, and the motivation to traffic cheaper wheat halfway across the world to feed England's industrial workers. At the other end of the California wheat trade in England, David Fouser in chapter 2 shows us how trade policies, class distinctions, and the visible appearance of bread swirled together to give shape to modern white bread. Rooted in the British imperial world, Fouser's biography of British bread shows that arguments over white and brown bread were also arguments over what was culturally better and worse. In contrast, as Alex Orquiza tells it in chapter 3, colonial control in the Philippines meant that moral categories of better or worse were not the result of hard-fought battles over the right kinds of food to eat, but the *cause* of them. Orquiza begins that tale with a puzzle from today: why isn't there a thriving Filipino-American cuisine in the same way one finds Chinese, Thai, or Vietnamese foods across the United States? To answer this, Orquiza unravels a story about the ways imperial measures to redefine Filipino lands led many white Americans to view the Philippines as a food culture to replace and exploit rather than understand, and in turn made Filipino people reluctant to share their cuisine with colonizers. Chapter 4 takes us back to Europe. Here, Jeffrey Pilcher tells of the invention of the first globally marketed beers, a triumph of new industrial production. Pilsner lagers found mass appeal because of a light, mild flavor that fit the new market demands of nineteenth century hygiene and purity, helping them become the first beer style to spread across the world. In the final chapter of part I, Tom Okie describes a modern kind of food distribution to think about how food gets from farm to table in order to limit rot and decay. His Great Watermelon Fiasco, with stevedores knee-deep in rotting southern fruit on the docks of Manhattan, explains how we came to manage death over time as a core element of modern distributed food.

Part II brings trust to the fore. Fake sugar, breakfast cereal, radical food stores, bananas, and canned food offer snapshots of contests over authority, moral and otherwise, amid changing relationships. Chapter 6 offers a story about fake sugar in an era of charlatans. In it, Benjamin Cohen shows that trusting alternative sweeteners like glucose or corn syrup depended on overcoming suspicions about coastal elites, large-scale environmentally intensive factories, and prevailing fears over hucksters and con men. During the same era, Michael Kideckel shows in chapter 7 how breakfast cereal, clearly new and sold as "natural," transformed American diets due to unprecedented marketing and rapid adoption. In wondering why an individual family may have trusted strange grains from cardboard boxes, Kideckel explores the promise and perils of seeking a typical consumer with which to understand the industrialized food supply, one cereal box at a time. The power to prescribe food habits was not just commercial but also institutional and racial. In chapter 8, Faron Levesque intertwines two characters and settings: Ida B. Wells and the People's Grocery in Memphis, and Lucy Parsons and Chicago Black liberation movements. Levesque highlights a new kind of advocate-activist working for health, well-being, and food aid in ways that place modern food justice efforts into a much longer and more durable past. In this case, the story is as much about distrust of food communities as it is about trusting your neighborhood grocer. Even as white mobs fought to destroy Black food sovereignty, Tashima Thomas describes in chapter 9 how racialized advertising used Black bodies to bring the appeal of foods to ever larger consumer masses, always rooted in cultural and gendered connotations. In Thomas's story of Josephine Baker's famous banana skirt, fruit, performance, sexual innuendo, humor, Blackness, and the colonial gaze were all bound up in the song and imagery of colonial fruits and their global marketing. Finally, Anna Zeide's story of famed domestic advisor Marion Harland in chapter 10 shows the origins and power of celebrity endorsement in garnering consumer trust. Harland's changing views on canned food helped give stability to an industry that moved from still-shaky foundations at the end of the nineteenth century to a central pillar of industrial foods by the early twentieth century, in part

because of Harland and the values of domestic science and professionaliza-
tion that she promoted.

Part III is organized around science to show how dislocation of physical
and moral relationships led to the triumph of institutional expertise, which
promoted efficiency, racial categories, and the idea that one food could
substitute for another. In chapter 11, David Singerman finds racially coded
forms of food provisioning in his story of cane sugar imports and scientific
analysis. Fears of invasion by workers of darker skin and foreign conspira-
tors motivated an approach to policing sugar quality on the docks and in
customs bureaus as a way of policing the purity of white American bodies.
Chapter 12 has Lisa Haushofer narrating the arguments of British imperialism to reduce and concentrate meat to feed the masses in a time of limited
resources. Their solution? The strange invention of "fluid meat." In the
late 1800s, what could have been an arcane debate between physiologists
and chemists over who got to define food became an intriguing combination of nutritional research, British imperial commitments, and scientific
insights on digestion. At the same time, we learn from Adam Shprintzen
in chapter 13 that modern American fake meat was born around 1900 in a
food laboratory in Battle Creek, Michigan. It was the result of years of work
by Ella Eaton Kellogg, the pioneer who has long been overshadowed by
her husband, John Harvey Kellogg. Amrys Williams moves the focus from
cities and ports to the northern Rockies of Depression-era Montana, in our
final chapter, illustrating through a girl named Marietta and her 4H Club
lamb how rural education linked the scientific feeding of animals with the
scientific feeding of humans. In feeding and raising her sheep in a modern,
scientific way, Marietta became the modern eater.

The essays in this book add up to something diverse rather than singular.
Modern food isn't one thing. Modern food, as our contributors show, is
racialized, colonized, globalized. It is reduced, standardized, suspicious,
and endorsed by celebrities. It is plural. And there are so many more stories of modern food that our collection barely touches on in its limited
scope—especially those from Latin America, Africa, and throughout the
Global South, from immigrant communities, from those displaced or reset-
tled, and from those outside the power structures that have sought to

develop and then impose an idea about what foods are best, relevant, and possible.[10]

New approaches to collapsing time and space, building trust, and developing scientific protocols created the basis for crafting what we, having moved our chairs up to the twenty-first century, think of as a modern food system. That usable past is what we see as we look ahead. Our contributors think of a better future as one that is healthier for the land, for our bodies, and for our cultural relationships. To help achieve that future, we need to develop ideas and forms of innovation based on overcoming the limitations of a damaging food environment brought to us by cultural changes larger than food alone. Policies about the future of food that do not attend to racist coding and exploitation, the consequences of colonial conquest, capital-intensive industrial agriculture, and mass marketing will only scratch the surface of a food system whose depths demand greater scrutiny. The limitations of our current food system are the result of developments, choices, and ideas formed over a century ago.

To help achieve that future, we also need to bring more stories to the fore, with storytelling that influences whether we think food has improved, become less trustworthy, or something in between. Telling stories about food and the environment has always been crucial, but is now more urgent in the face of the twinned crises of climate change and the problems of injustice and inequity plaguing the food system. Good stories, on their own, may not save us, but they give us a way to humanize our work, to emphasize historical characters and circumstances, and to show how individuals can change systems. We hope that the following narratives will engage students, scholars, food enthusiasts, food policy readers, and local food advocates who are eager for more history, and for a better future.

NOTES

1. Some starting points on history and writing include William Cronon, "A Place for Stories: Nature, History, and Narrative," *Journal of American History* 78, no. 4 (March 1992): 1347–1376; Cronon, "Storytelling," *American Historical Review* 118, no. 1 (February 2013): 1–19; James E. Goodman, "For the Love of Stories," *Reviews in American History* 26, no. 1 (1998): 255–274; Jill Lepore, "Historians Who Love Too Much: Reflections on Microhistory and Biography," *Journal of American History* 88, no. 1 (2001): 129–144; and the roundtable starting with Aaron Sachs, "Letters to a Tenured

Historian: Imagining History as Creative Nonfiction—or Maybe Even Poetry," *Rethinking History* 14, no. 1 (March 2010): 5–38.

2. Aristotle, *The Poetics of Aristotle*, trans. S. H. Butcher (New York: Macmillan, 1895), retrieved from http://classics.mit.edu/Aristotle/poetics.1.1.html.

3. Melanie Warner, *Pandora's Lunchbox* (New York: Scribner, 2014), xv; Katherine Gustafson, *Change Comes to Dinner* (New York: Griffin, 2012), 3; Michael Moss, *Salt, Sugar, Fat: How the Food Giants Hooked Us* (New York: Random House, 2013), xv. "In the universe of processed food, World War II was the Big Bang." Anastacia Marx de Salcedo, *Combat-Ready Kitchen: How the U.S. Military Shapes the Way You Eat* (New York: Penguin, 2015).

4. To be sure, writers asking about origins look to that period in part because they are diagnosing the real problems of diet-related diseases, processed foods, highly refined sugars and fats, and insidious marketing, not to mention the even deeper issues of hunger and injustice. Examples of this abound, but a few can be found in Bee Wilson, *The Way We Eat Now: How the Food Revolution Has Transformed Our Lives, Our Bodies, and Our World* (New York: Basic Books, 2019); Marion Nestle, *Unsavory Truth: How Food Companies Skew the Science of What We Eat* (New York: Basic Books, 2018); Mark Schatzker, *The Dorito Effect: The Surprising New Truth about Food and Flavor* (New York: Simon & Schuster, 2016); Michael Pollan, *The Omnivore's Dilemma: A Natural History of Four Meals* (New York: Penguin Press, 2006); Eric Schlosser, *Fast Food Nation: The Dark Side of the All-American Meal* (Boston: Houghton Mifflin, 2001); Moss, *Salt, Sugar, Fat*.

5. The Pollan-critique industry is nearly as full as the industries Pollan critiques. For a sampling, see Charles C. Ludington and Matthew Morse Booker, eds., *Food Fights: How History Matters to Contemporary Food Debates* (Chapel Hill: University of North Carolina Press, 2019); Maria McGrath, *Food for Dissent: Natural Foods and the Consumer Counterculture since the 1960s* (Amherst: University of Massachusetts Press, 2019), 197; Tracey Deutsch, "Memories of Mothers in the Kitchen: Local Foods, History, and Women's Work," *Radical History Review* 110 (2011): 167–177; and Michael D. Wise and Jennifer Jensen Wallach, eds., *The Routledge History of American Foodways* (Abingdon, UK: Routledge, 2016), 4 and 5.

6. Michael Pollan, "Six Rules for Eating Wisely," *Time*, June 12, 2006, 97; Pollan, *In Defense of Food: An Eater's Manifesto* (New York: Penguin, 2008), 148.

7. John McNeill and Matthew Engelke, *The Great Acceleration: An Environmental History of the Anthropocene since 1945* (Cambridge, MA: Harvard University Press, 2014), provides an entrée into the topic.

8. Warren Belasco, "Food History as a Field," in *Food in Time and Place: The American Historical Association Companion to Food History*, ed. Paul Freedman, Joyce Chaplin, and Ken Albala (Berkeley: University of California Press, 2014), 11; Laura Shapiro, *Perfection Salad: Women and Cooking at the Turn of the Century* (New York: Farrar, Straus, & Giroux, 1986), 201. For a view against "culinary Luddism," see Rachel Laudan, "A Plea for Culinary Modernism: Why We Should Love Fast, Modern, Processed Food (With a New Postscript)," in *Food Fights: How History Matters to*

Contemporary Food Debates, ed. Charles C. Ludington and Matthew Morse Booker (Chapel Hill: University of North Carolina Press, 2019), 262–284. See also James C. Giesen and Mark Hersey, "The New Environmental Politics and Its Antecedents: Lessons from the Early Twentieth Century South," *The Historian* 72, no. 2 (Summer 2010): 271–298; and Nathan A. Rosenberg and Bryce Wilson Stucki, "The Butz Stops Here: Why the Food Movement Needs to Rethink Agricultural History," *Journal of Food Law and Policy* 13, no. 1 (Spring 2017): 12–25. See also Charlotte Biltekoff, *Eating Right in America: The Cultural Politics of Food and Health* (Durham, NC: Duke University Press, 2013); Michael S. Kideckel, "Anti-Intellectualism and Natural Food: The Shared Language of Industry and Activists in America since 1830," *Gastronomica* 18, no. 1 (Spring 2018): 44–54. For a good example of the existing scholarship on the period of the later 1800s and early 1900s, see "Food Studies and the Gilded Age and Progressive Era," special issue, *Journal of the Gilded Age and Progressive Era* 18, no. 4 (October 2019).

9. William Cronon wrote of such "declensionist" structures some time ago in "A Place for Stories"; see also Paul S. Sutter, "The World with Us: The State of American Environmental History," *Journal of American History* 100, no. 1 (June 2013): 94–119.

10. Many have told those stories and we urge readers to engage with their narratives. To give just a sampling, see Rebekah E. Pite, *Creating a Common Table: Doña Petrona, Women, and Food* (Chapel Hill: University of North Carolina Press, 2013); Candace Goucher, *Congotay! Congotay! A Global History of Caribbean Food* (New York: Routledge, 2015); Matt Garcia, Melanie DuPuis, and Don Mitchell, eds., *Food across Borders* (New Brunswick, NJ: Rutgers University Press, 2017); Devon Peña, et al., eds., *Mexican-Origin Foods, Foodways, and Social Movements: Decolonial Perspectives* (Fayetteville: University of Arkansas Press, 2017); Kyla Wazana Tompkins, *Racial Indigestion: Eating Bodies in the Nineteenth Century* (New York: New York University Press, 2012); Jennifer Jensen Wallach, ed., *Dethroning the Deceitful Porkchop: Rethinking African American Foodways from Slavery to Obama* (Fayetteville: University of Arkansas Press, 2015); Elizabeth Zanoni, *Migrant Marketplaces: Food and Italians in North and South America* (Champaign: University of Illinois Press, 2018); Michael Twitty, *The Cooking Gene: A Journey through African American Culinary History in the Old South* (New York: HarperCollins, 2018).

I

TIME AND SPACE

1

Modern Food as Ghost Acres

TULARE LAKE AND THE PAST FUTURE OF FOOD

Thomas D. Finger

If you open Google Earth and zoom in between the towns of Corcoran and Kettleman City, California, you will find a ghost. Shrouded by fields of soybeans, alfalfa, and corn, Tulare Lake lies hidden. What was once the largest freshwater lake west of the Mississippi River is now rows of grain crops and vegetables. These crops feed millions of people across the United States and the Pacific world on the stored wealth of minerals locked deep in the crust, water melting high in the mountains, and the ecosystem wealth built from generations of management by the Yokuts people of California's Central Valley (figure 1.1).

Tulare's history has often been told as a water story.[1] This chapter explores its food history. During the second half of the nineteenth century, an abundant landscape of tule marshes and oak stands that fed Yokuts peoples transformed into a giant wheat field feeding industrialization in England. This involved human violence as well as ecological collapse, creating a landscape that historian Kenneth Pomeranz refers to evocatively as "ghost acres."[2] One of the many important tasks of food history is to provide an account of the creation of ghost acres, of asking how they come together and break apart in the hope of providing some answers about how healthy communities may reform in the future. Ultimately, food histories of ghost acres must revolve around the changing relationships among human communities across time, wrestling with the

1.1 *Map of the San Joaquin, Sacramento and Tulare Valleys.* Tulare Lake lies at the southern end of California's Central Valley. Note the numerous feeding streams. These streams fed the lake with spring snowmelt from surrounding mountains, contributed to the ecological patchwork of the area, provided the Yokuts people with a diverse diet, and were deepened and straightened into irrigation canals by white settlers after the 1850s. By the 1870s and 1880s, much of the Central Valley bottomland was planted in wheat exported to England from San Francisco. (California Irrigation Commission, 1873, David Rumsey Historical Map Collection.)

tangle of historical memory while envisioning a just future for those who continue to live in—or are fed by—landscapes hijacked during the creation of global food flows.

Ghost acres are not dead; they are in a state of suspended animation. Today, Tulare Lake supports water-intensive, high-fertilizer industries, in freeze-dried vegetables, soy, and cattle, all of which pump overworked food into overfed stomachs. Places around the world underwent a similar process as Tulare, transforming from a complex seasonal network of food production to uniform suppliers of western capitalism. Understanding those histories can help guide the ecological and human reanimation of ghost acres.

Modern food is not just the products on the shelves, but the acres transformed so we can consume those products. The nineteenth century laid the groundwork for those new places and Tulare Lake, a ghostly landscape, is an example of the violence written into such locales. Mostly, this now-hidden lake in one of the world's most productive agricultural regions helps us think about other unacknowledged portions of our food system. The Yokuts of California's Central Valley still by and large live in their homeland on the Santa Rosa Rancheria near the ghost lake of Tulare. Many are employed by the tribal government or the large Tachi Palace Casino and Hotel just outside Lemoore, California. Each jar of peaches or glass of milk produced on Tulare bottomland owes a portion of its nutrients to the land and water management of the ancestors of those who live and work almost within sight of the dried lake. Just as their nation's history was wrapped up in the gradual disappearance of the lake and in grander schemes to collapse time and space, so too can their future be further enriched by its return.[3]

Let me take you now to Ciau, a village just north of Tulare Lake in 1840. The Nutunutu band of the Yokuts people migrated around their central village, making use of the valley's rich ecological diversity and seasonal patterns of productivity to forge a healthy and resilient diet. Nearby, a seasonal channel later called Mussel Slough moved water back and forth between the Kings River and Tulare Lake. Here, Yokuts women migrated to find food according to the pulses of seasons.[4]

The women of Ciau understood this was a place of dramatic variation in water and temperature. Prior to the late nineteenth century, the San Joaquin Valley surrounding Tulare Lake was a patchwork of niches, running the ecological gamut from alkaline desert to lush riparian forest. The western portions of the Tulare Basin lay in the rain shadow of coastal mountain ranges and were the most arid regions of the valley. Moving further east, alkaline flats gave way to short grasslands. These grasslands in turn transitioned to salt marshes that thrived along the wavering shoreline of Tulare Lake. Tall grasslands predominated as the land rose from lake to foothills, on which thrived oak parkland. Tendrils of sycamore forest clung to waterways which cut toward the lake.[5]

Tulare Lake itself filled entirely from snowmelt in the Sierra Nevada Mountains each spring, and would then shrink considerably in the hot, dry summer months. Dimensions of the lake changed dramatically from season to season, and year to year. The lake measured only forty feet deep at its deepest location, and in many places it was shallow enough for someone to wade out a mile up to their shoulders. In wetter years, Tulare could cover as much as 1,800 square miles. Early surveys record the area around the lake as "overflow and swamp," and early white settlers recorded extremely wet years in 1852, 1858, 1860, and 1868. Despite this, the lake almost dried completely in 1854. One local newspaper reported in 1890 that the lake had spread out a full fifteen miles during the spring season at a rate of about a mile and a half per week. Another early settler remembered that "the shore lines of Tulare lake changed and shifted a great deal. If a strong wind came from the north, as it often did, the water would move several miles south, and would move again when the wind changed. Then when the water level in the lake changed the shore line shifted a long distance." This hydraulic metabolism created a rich diversity of ecological niches surrounding the lake, including Mussel Slough and nearby Ciau.[6]

The rich environments of the Tulare region produced what ecologists call the portfolio effect: no matter what the season or weather conditions, some resource was experiencing peak productivity. The area must have been beautiful, and certainly was bountiful. One early white settler recalled wistfully,

I have always remembered that place as one of the most ideal I have ever seen. The tall green grass, the cool clear water, and the trees with their fresh leaves made as pretty a sport as one could wish . . . for my own part, I have never seen anything equal to the virgin San Joaquin Valley before there was a plow or a fence in it.[7]

Before the latter half of the nineteenth century, we know the inhabitants of Ciau rarely—if ever—went hungry. As one early settler history of Tulare County recounts, "Acorns, of course, were the staple, but it is a mistake to suppose that the Indians' diet lacked variety. In addition to game of all kinds and fish, there were various kinds of seeds, nuts, berries, roots, and young shoots of the tule and clover." There were once actually mussels in Mussel Slough, which the Nutunutu collected and boiled in fine-woven baskets filled with hot stones.[8]

As temperatures cooled in the fall, Yokuts women left Ciau for their seasonal camp in the oak-studded foothills some forty miles to the east. Here, acorns dropped from oaks fertilized by fires delicately managed by the village's men a few months prior. Women would collect piles of acorns at the creekside and pound them into a rough flour using a large mortar and pestle fashioned from a tree trunk and a large branch before pounding the flour into a cake to let it dry on a rock.[9]

With abundant food, there was little open conflict between the Tachi, Nutunutu, and many other Yokuts bands migrating around their central place. Then came consecutive blows to the Nutunutu and larger Yokuts communities in the first half of the nineteenth century. First was the expansion of the Spanish mission chain up the Pacific Coast. Because Ciau sat well inland, the major impact from the missions was the introduction of grazing animals, whose ranging contributed to the collapse of native grasses and the introduction of foreign species. As Spanish and Mexican cattle herds fanned out across the valley, native plants died and the Nutunutu saw some strands of their dietary network collapse. Throughout the early nineteenth century, they, like many other Yokuts bands, made up for the loss of native species by adopting the horse to ride, raid, and eat. Even as they made these adaptations, other forces conspired. In the early 1830s, malaria swept in, carried by traders for the Hudson's Bay Company. Finally, in the 1840s came the cataclysm of gold mining and Anglo settlement. The next half century would see the remaking of this place as a depot for global trade, a place where violence shook food systems that had before been marked by self-reliant sustenance.[10]

As the Nutunutu found their dietary network stressed, halfway across the world so too did the poor workers of Manchester, England. The two communities eventually became linked by the flow of wheat grown on stolen Nutunutu land in California and shipped overseas to feed the industrial English workers with cheap bread. Manchester qualified as a village at the beginning of the 1700s and was only given representation in Parliament in 1832. But as the Nutunutu reeled from disease, Manchester became a fast-growing city of shanties, warehouses, and manufactories. The city grew as forces of land dispossession and food commercialization forced people to migrate from farms to cities. It was a city of hungry refugees.[11]

Across England in the nineteenth century, many working-class women felt the daily brunt of these systemic changes to their food economy. Subsistence landscapes transformed into commercial farms; growers became eaters. The life of an English rural family prior to being kicked off their land would have been seasonal in nature, not terribly unlike that of the Nutunutu. Their food came from fallow-rotation farming, fishing in local creeks, hunting game, and foraging herbs and root vegetables in forests, fens, and unplowed fields. But accelerating enclosure consolidated land rights in the hands of a country gentry. As had happened to the Nutunutu, English rural methods of obtaining food began to crumble. Long rows of turnips, barley, and oats turned into geometric wheat fields. Forests were felled and plowed into fields. Experts drained fens, devised new labor-saving machines, and innovated fallow patterns designed to feed cities bulging with the very people who had once worked the land.[12]

Just like the Nutunutu, unsettled English people in the nineteenth century moved around a central place. Crucially, though, this was a migration borne of desperation, not seasonal forage. Most would come to work in burgeoning towns like Leeds, Birmingham, or Manchester, but hunger and housing prices kept them ever on the move. There was a strongly gendered component to the situation too. Women were expected to run the household, which included the ever-taxing work of finding and cooking food. They were also often called to supplement the family's income through wage work. These industrial settlements were places of work opportunity, but little dietary resiliency, as people worked all day in food-less environments. There was little room for gardens in tightly packed neighborhoods. Women often bore the brunt of these problems: not only would they shoulder the stress of obtaining and cooking enough food for their families (and anyone else who might be living in their crowded quarters), they were often expected to be the first to forgo food in times of shortage, leaving sufficient food for the male "breadwinners" and their developing children, and supplementing their diets with energy-packed, nutrient-deficient sips of tea and bites of candy. Women's mental and physical health suffered.[13]

This situation reached a crisis point in the 1840s. For so many English working-class women in the nineteenth century, universal suffrage—and politics in general—were "a knife and fork question," about eroding the power of the landowning class who, through enclosure, had simultane-ously cast people from traditional lands and made them dependent on the whims of the market. Merchants and laborers of newly industrial cit-ies rose to the cry of suffrage and cheap bread. Atmospheric conditions over the North Atlantic pumped a depressing slate of cool, wet weather over northern Europe in the early 1840s. Crops failed. Families went hun-gry everywhere in the British Isles and northern Europe. While the "hun-gry forties" is most well known from the Irish Potato Famine, the poor all over Europe found their dietary stress dip into outright starvation. And so, England—and the poor of Manchester—cast a hungry eye out across the world.[14]

At the same time, Yokuts women noticed strangers in their oak stands. American settlers in the 1840s and 1850s preferred to settle in the oak-blanketed foothills of the eastern slopes of the Tulare region not because of the acorns they contained, but as refuge from the floods and seasonal swampland that characterized much of the valley's bottom. Here, set-tlers began planting wheat and alfalfa, girdling or cutting down so many of the acorn trees those of Ciau had worked so hard to keep. Much of this crop went to feed magnate Henry Miller's vast cattle empire, sprawl-ing as it was across the central portions of the valley. These crops would then be consumed by cattle and cattle workers and, when the cattle were slaughtered and butchered, would feed hungry miners toiling away on busted claims high in the mountains.[15]

Later arrivals moved into the marshlands below. Businessmen lurked for profit and new passions enflamed old land disputes. Failed miners and desperate emigrants rushed to the Tulare region to make quick cash from cheap native land, often just given away by the new state government. What couldn't be taken legally was taken by force.

In a story of violent place-making that repeated itself thousands of times across the globe in the nineteenth century, newly arrived white farmers in the Tulare region went to war with the remnants of Ciau and other Yokuts communities in 1856. Farmers quickly formed a local militia group after drumming up an excuse, intent on asserting their settler campaign. The few remaining Yokuts living in remnant villages throughout the Tulare region rightly saw that preparations would lead to a punitive expedition. They headed into the foothills for protection. Ciau emptied, never to be filled again.

The militia group chased the refugee community to the foothills and sought to dislodge them. Their attack in May 1856 was at first unsuccessful, but aid from a federal cavalry unit stationed just south of the valley ultimately helped them overwhelm the Yokuts, sending survivors scattering into the mountains. For weeks after, the militia and federal cavalry rode across the countryside, alternatively killing or gathering any native person they found and incarcerating the survivors at the Kings River Indian Reservation just south of Fresno. And so, through the cumulative force of disease, environmental change, and state violence, the landscape of Tulare became unmoored in the global economy, its people killed or captured.[16]

Into this scene stepped Isaac Friedlander, a six-foot-seven, three-hundred-pound German immigrant whose shadow would loom large over the Tulare Valley for the next twenty years. Friedlander at first was as unmoored as the Tulare landscape he would come to own. After a series of failed business ventures on the East Coast, he came to California in 1849 hoping to strike it rich. Here he failed again and tumbled back to San Francisco in 1852 to reevaluate his business prospects. He read reports streaming back from the interior of "wars" that were clearing the Yokuts, Wappos, Miwoks, and

Patwins from their lands. He also happened to meet two very important businessmen, banker William Ralston and land lawyer William Chapman. Together, Ralston, Chapman, and Friedlander hatched one of the most crooked land schemes in American history. Their plan sowed the seeds of Tulare Lake's demise by connecting it to Manchester. No longer would the food ecology of Tulare be controlled by temperature and moisture, but by account balances and interfirm competition. Ralston bankrolled the operation with gold-backed securities from his Bank of California; Chapman massaged the law. With these partners, Friedlander looked hungrily to the southern valley and the Tulare region. He obtained his land in several ethically dubious or outright illegal ways. By the late 1850s you might have seen Friedlander waylaying drunken patrons of the local saloons, plying them with a scheme: Friedlander would ask them to accompany him to the public land office. While land law limited the amount of dispossessed land one person could buy from the state, Friedlander would slip the man enough money to make a purchase and then immediately buy it back from him once he stumbled out. The next day, you might find Friedlander in his office at the University of California, using his membership on the Board of Regents to privately buy and sell land nominally reserved for public schools. Finally, he'd walk over to the state land surveyor's office and whisper into the ear of one Isaac Chapman, who just happened to be the brother of the land lawyer. Isaac Chapman would then declare some portion of the Tulare basin to be "swamp or marshland," and Friedlander would use Ralston's credit to buy it for pennies on the dollar. To hide these spurious land deals, Friedlander and his associates would resell the land in various parcels back and forth to each other, burying the original deal in a mountain of paper in what could only be described as a land-laundering operation. Through such means, Friedlander came to own 500,000 acres in the Central Valley by the late 1860s.[17]

As Friedlander and his cronies bought and sold land, small-time settlers—many of them migrants from the American South—came to occupy it. These settlers set to work transforming Yokuts land into wheat fields, producing wheat destined for the global food market. It was their ditches and wheat plants that would eventually suck up water that normally coursed down from the mountains and settled in Tulare Lake. The seeds of the lake's destruction were being planted right along Mussel Slough and beside the ghost of Ciau.[18]

Here, significant ecological transformation was accompanied by continued human tragedy. The land around Mussel Slough had been taken so quickly, so lawlessly, that it was bound in a jumble of conflicting claims. Friedlander and associates, cattle magnate Henry Miller, railroad companies, and squatters all claimed some portion of land and water rights in the area. These overlapping claims created bizarre scenes and comical cat-and-mouse games. Dry farmers who planted wheat were never sure they would hold onto the land long enough to make improvements profitable. Often, they simply stored their grain in whatever structure they managed to build and slept in tents; others stored sacks out in the open, praying that the rain would hold off. Families formed mutual aid societies that often took the form of armed vigilante groups roaming the countryside, protecting whatever right they believed they were owed. Railroad agents scoured the countryside serving eviction notices. The families would often hide at their approach so that these agents, finding empty houses, would place all the furniture outside and padlock the house. Hiding families returned to break the lock and replace the furniture.[19]

People continued to die for land along Mussel Slough in the 1870s and 1880s. Vigilante groups became more militant. Railroads locked up claims in court battles. The situation escalated until, in 1880, seven people died when one of these ridiculous land squabbles turned violent as gunfire erupted in a wheat field. What only a generation before was a verdant landscape tightly ordered by customary rights was now a landscape of disorder and violence increasingly controlled by a syndicate of conmen and wracked by gang violence.[20]

The white settlers were not acting in a vacuum. Rather, they operated in a global drama that not only found them agents in an expansionist agenda for the country but tied them to a continent an ocean away. As white settlers transformed Mussel Slough, British factory workers looked hungrily around the world for their own daily bread, a story that David Fouser alludes to in chapter 2. Cities like Manchester were simply growing too fast for domestic farmers to feed them. At the same time, Friedlander came to own too much wheat. He had peopled his vast estates with tenant farmers who paid their rents in wheat. By the 1860s, Friedlander was holding so much that he could set prices across the entire valley, so much that hungry miners in the Sierra and sailors in San Francisco Bay couldn't eat it all. Friedlander sought to offload the sullied bounty by chartering ships to sell his wheat to whoever would buy it. His first attempt had the ships docking in far-flung settlements across the Pacific world: Oregon, Hawaii, the Philippines, and Chile. But there was so much that such an ad hoc approach could never sell it all. The hungry markets of England started to look appealing. Friedlander began traveling every other year to Liverpool to set up business deals for his California wheat. For years until his death in 1878, Friedlander would send a "grain fleet" of 200–400 ships to England laden with wheat grown alongside Mussel Slough and countless other creeks across the Central Valley.[21]

By then, the Tulare region—and the larger Central Valley itself—was one vast field of wheat that stretched from coastal ranges to the Sierra Nevada foothills. Its new shape began to look like the modern world's epitome of agricultural productivity. Wheat was pumped out of the valley in railroad cars and steamship holds where sailors loaded bags onto bulky sailing vessels. The wheat would breathe with heat and cold as it crossed the tropics, hit the Antarctic Cape Horn passage, reenter the tropics, and cross into the North Atlantic. After the 14,000-mile journey it would enter the gigantic port of Liverpool before being milled into flour and eventually sold. Some of that flour found its way to a small baker in downtown Manchester. It was there, sitting on a shelf of a Manchester bakery, that the bread would wait for a tired woman to enter, buy it, and carry it home to her hungry family for dinner. Through this journey that would repeat itself thousands of times during the second half of the nineteenth century, Tulare now fed Manchester.[22]

The world Friedlander helped create lived beyond his death. By the 1880s, the Manchester working-class family was no longer as hungry as their parents and grandparents had been in the 1840s, thanks in large part to Friedlander, but they were just as unhealthy. Their new diet was high in energy but nutrient deficient. What's more, their sapped immune systems had to live surrounded by the digestive tracts of thousands of other people. The global system engendered by developments in California and places like it meant that these people found just enough food at regular intervals to finally settle down. The European industrial class no longer had to migrate constantly to find food and work; they could simply walk to the corner bakery. Bacterial diseases exploded as human waste flooded inadequate privies and sewers.[23]

Manchester's eating patterns also helped destroy Tulare Lake. The next century saw an acceleration of the patterns of land use developed around Tulare. White settlement, for instance, accelerated when land titles became clear in the valley, with the best land snatched up before the end of the 1800s. Many new settlers shut out from lucrative valley plots bought land in the foothills—once the center of the Yokuts' acorn-based food system. They grazed sheep on grasses between now-untended oak trees. When deep droughts destroyed these animals, some desperate farmers began planting orange trees. By the turn of the century, irrigation ditches snaked across foothills and laced valley bottomland, all feeding plants that would in turn feed humans thousands of miles away.[24]

Tulare Lake itself began to buckle under the pressure of so much withdrawal. A new assault on Tulare Lake came as fishmongers in San Francisco looked hungrily south. The lake sat relatively undisturbed until the spring of 1884. In that year, the *Tulare Daily Register* reported, "The rivers ran into the lake in a flood that spring and the fish met the fresh water in solid masses. Standing at the mouth of King's river, one could see a wave come landward, a wave produced by the motion of a mighty army of fish. . . . The ditches of Mussel Slough country were choked by them." Fishermen swarmed so quickly that the Tulare fishery was dead by the early 1890s.[25]

An irrigation craze swept over the Tulare district toward the end of the century as land titles cleared and global demand for wheat and oranges soared. Farmers began to dream of reclaiming the area of Tulare Lake into waving fields of wheat, a dream that was growing into reality when the Buena Vista reclamation levee was completed in early 1899. Its builders hoped to repossess 30,000 acres of Tulare bottomland and seed them all with grain. Soon cotton, melon, and alfalfa joined wheat to grow on what was once lake bottomland. By the 1920s, over a million bags of grain and 15,000 bales of cotton were collected annually from the dead lake.[26]

But by then, Tulare Lake was cut off from spring snowmelt and no longer breathing with a hydraulic metabolism; it was suffocating. The local paper warned, "Tulare Lake Disappearing: large and fertile farms where lake once stood. Tulare Lake is drying up. Its waters are constantly receding. Like the dawning of a new creation, pleasant groves and fertile fields take the place of its former wastes of waters." Another paper reported that "Tulare Lake is as dry as a chip. For the first time in recent history, the pelican, geese, ducks, snipes, mud hens . . . as well as the many fish have found that there is no longer a home for them." Looking back with quick retrospection in the 1930s, the *Fresno Bee* wrote, "The lake has only been a puddle since 1920, having not really existed since 1917 and not having had its customary size since 1906."[27]

Ecological collapse was everywhere. The disappearance of Ciau and the murder of Yokuts people laid the groundwork for a new world of commodities. It hadn't happened by chance. Migrating settlers, Friedlander's crooked land scheme, and wheat purchases by English millers were all predicated on one another. Each decision was made with a fleeting understanding of what was happening elsewhere. Settlers and capitalists often emphasized how they made "improvements" to nature with a seemingly rational design of food economy. But the disappearance of Tulare Lake told another story, one connected to the plight of surviving Yokuts held on reservations. With the deck cleared by midcentury, agents, fishermen, irrigators, and levee builders dealt the lake a final blow. A new, spectral place—these thousands of ghost acres—Tulare Lake traded death for life. The bread that so many relied upon came from the demise of people, of landscape, and of nutrients. Their histories cannot be unbound.

Tulare no longer feeds Manchester, but American and Asian supermarkets. You've likely consumed a part of that place. As of 2018, the region was among the nation's largest producers of dairy products, fruit, and cotton. All this production grows from generations of genocide, land dispossession, and ecological collapse. The Yokuts people have witnessed it all. The Tachi Yokuts of the Santa Rosa Rancheria endured in the face of such momentous change. The land—and its ghost lake—occupies a central place in their culture and memory. As tribal historian Raymond Jeff recounts, settlers "killed the whole San Joaquin valley." "I've never even seen the lake," he laments, "all I did was read about it." Importantly, however, the future of this story can be one of rebirth, not death. As the tribe proclaims, "Now, we rebuild. We will endure."[28]

The ghosts that lie within our shared food history can guide us to the future. Let the water flow. The water will come back, has come back. In the 1930s, persistent floods allowed nesting pelicans to return briefly to the lake. Such is but a glimpse of the reanimation that can unfold in erstwhile ghost acres. Part of that restoration must include the Yokuts people and their traditional foodways. The reanimation of ghost acres does not only call for nature preserves or wildlife parks. If our food history played such a role in arresting complex ecosystems like the Tulare Basin, so too must our eating patterns help reanimate them. Breathing new life into our food system should revolve around restoring Indigenous stewardship and decision-making over their food landscapes. As ethnobotanist Kat Anderson notes, "the conservation of endangered species and the restoration of historic ecosystems might require the reintroduction of careful human stewardship rather than simple hands-off preservation." This "Indigenous management model" can not only restore vibrant cultural food landscapes, it can be productive as well. Modern studies are beginning to appreciate that Indigenous land management was significantly more productive than originally believed by Western science. It is time to use past models and current knowledge to replace food systems built on genocide and land mining.[29]

The reanimation of the ghost acres of Tulare Lake will require policy that acknowledges Indigenous claims to land and water and the academic expertise of ethnobotanists, food historians, agronomists, and ecologists— but it should be led by the Yokuts people. Only they have proven adequate to the task of feeding large numbers of people while at the same time storing nutrients for future use. Food is an excellent venue through which to build public university–community collaborations. Due to the traumatic history, only hard-earned trust and embedded relationships can restore food landscapes that acknowledge past trauma and use Indigenous and academic knowledge to reanimate ghost acres into verdant landscapes that will feed us in the future.[30]

Truly restorative food justice will use history and dialogue with people like the Yokuts, who still live next to ghost acres held back by irrigation canals, herbicides, and plows. Let the deep wisdom held in place by Yokuts past and present seep into small networks of irrigation canals that water a patchwork landscape of local plants and agricultural plots tended by scientific knowledge. History, restorative ecology, and food culture studies can work together to create just food systems that work within the patterns of place. Only then will the ghost of Tulare Lake truly come alive.

NOTES

1. For examples of this water emphasis, see Donald Pisani, *From the Family Farm to Agribusiness: The Irrigation Crusade in California and the West, 1850–1931* (Berkeley: University of California Press, 1984); Marc Reisner, *Cadillac Desert: The American West and Its Disappearing Water,* rev. and updated ed. (New York: Penguin Books, 1993).

2. Kenneth Pomeranz, *The Great Divergence: China, Europe, and the Making of the Modern World Economy* (Princeton, NJ: Princeton University Press, 2000), 201–202.

3. U.S. Census Bureau, "My Tribal Area," https://www.census.gov/tribal/?aianihh=3520, accessed July 15, 2020; Tachi Yokut Tribe, "Welcome to Tachi Yokut," https://www.tachi-yokut-nsn.gov/, accessed July 15, 2020.

4. The term "Yokuts" refers to a large linguistic and cultural group. Village units comprised of interwoven networks of families and clans made up the primary Yokuts group structure. Historically, the Tachi and Nutunutu Yokuts lived near each other and shared resources. This shared resource use unfolded mostly in harmony but also sometimes with flashes of disagreement. Allied together in the face of Spanish incursion, these villages increasingly came to be grouped together as "Yokuts" by outsiders. A. H. Gayton, *Yokuts and Western Mono Ethnography I: Tulare Lake, Southern Valley, and Central Foothill Yokuts,* Anthropological Records 10:1 (Berkeley: University of California Press, 1948); Lisbeth Haas, *Saints and Citizens: Indigenous Histories of Colonial Missions and Mexican California* (Berkeley: University of California Press, 2014).

5. William L. Preston, *Vanishing Landscapes: Land and Life in the Tulare Lake Basin* (Berkeley: University of California Press, 1981), 4–29.

6. *Fresno Republican,* December 1, 1929; "1932 Runoff," Box 99, Folder 6, and "Tulare Lake—Drying Up in 1898," Box 99, Folder 9, Frank F. Latta Collection, Skyfarming: Finding Aid, Huntington Library, San Marino, CA (hereafter Latta Collection); F. F. Latta, *Uncle Jeff's Story: A Tale of a San Joaquin Valley Pioneer and His Life with the Yokuts Indians* (Tulare, CA: Press of the Tulare Times, 1929), 31.

7. Latta, *Uncle Jeff's Story,* 9.

8. Brian M. Fagan, *Before California: An Archaeologist Looks at Our Earliest Inhabitants* (Walnut Creek, CA: AltaMira Press, 2003), 290–292; Eugene L. Menefee and Fred A. Dodge, *History of Tulare and Kings Counties, California, with Biographical Sketches of the Leading Men and Women of the Counties Who Have Been Identified with Their Growth and Development from the Early Days to the Present* (Los Angeles: Historic Record Co., 1913), 119; Latta, *Uncle Jeff's Story,* 23.

9. Preston, *Vanishing Landscapes,* 31–46; Brooke S. Arkush, "Yokuts Trade Networks and Native Culture Change in Central and Eastern California," *Ethnohistory* 40, no. 4 (1993): 619–640.

10. Haas, *Saints and Citizens*; John Ryan Fischer, *Cattle Colonialism: An Environmental History of the Conquest of California and Hawai'i* (Chapel Hill: University of North Carolina Press, 2015); Andrew C. Isenberg, *Mining California: An Ecological History* (New York: Hill and Wang, 2005); Brendan C. Lindsay, *Murder State: California's Native American Genocide 1846–1873* (Lincoln: University of Nebraska Press, 2015); Benjamin Madley, *An American Genocide: The United States and the California Indian Catastrophe, 1846–1873* (New Haven, CT: Yale University Press, 2017).

11. P. Laslett, *The World We Have Lost* (New York: Charles Scribner's Sons, 1965); Robert C. Allen, *Enclosure and the Yeoman* (Oxford: Clarendon Press, 1992); Joan Thirsk, *Food in Early Modern England: Phases, Fads, Fashions 1500–1760* (London: Hambledon Continuum, 2006).

12. Allen, *Enclosure and the Yeoman*; J. D. Chambers and G. E. Mingay, *The Agricultural Revolution, 1750–1880* (London: B. T. Batsford, 1966); Mark Overton, *Agricultural Revolution in England: The Transformation of the Agrarian Economy, 1500–1850* (Cambridge: Cambridge University Press, 1996).

13. For a selection of relevant works, see Asa Briggs, *Victorian Cities* (Berkeley: University of California Press, 1963); John Bohstedt, *The Politics of Provisions: Food Riots, Moral Economy, and Market Transition in England, c. 1550–1850* (Farnham, UK: Ashgate, 2010); Patricia Fumerton, *Unsettled: The Culture of Mobility and the Working Poor in Early Modern England* (Chicago: University of Chicago Press, 2006); Friedrich Engels, *The Condition of the Working Class in England.* (Stanford, CA: Stanford University Press, 1968); Edward Smith, *Practical Dietary for Families, Schools, and the Labouring Classes* (London: Walton and Maberly, 1865), 198–201; B. Seebohm Rowntree, *Poverty: A Study of Town Life*, 4th ed. (London: Macmillan & Co., 1902).

14. *Northern Star*, September 29, 1838; n.a., *The Annual Register, or a View of the History and Politics of the Year 1838* (London: J. G. F. Rivington, 1839), 311; Thomas Carlyle, *Chartism*, 2nd ed. (London: James Fraser, 1840); A. Briggs, *Chartist Studies* (London: Macmillan and Co., 1960); Richard Cornes, "Early Meteorological Data from London and Paris: Extending the North Atlantic Oscillation Series" (PhD diss., University of East Anglia, 2010); Charlotte Boyce, "Representing the 'Hungry Forties' in Images and Verse: The Politics of Hunger in Early-Victorian Illustrated Periodicals," *Victorian Literature and Culture* 40 (2012): 421–449.

15. John Ludeke, "The No Fence Law of 1874: Victory for San Joaquin Valley Farmers," *California History* 59, no. 2 (Summer 1980): 98–115; David Igler, *Industrial Cowboys: Miller & Lux and the Transformation of the Far West, 1850–1920* (Berkeley: University of California Press, 2001).

16. Menefee and Dodge, *History of Tulare and Kings Counties, California*, 20; Madley, *An American Genocide*, 244–246.

17. "The California Grain King: Sketch of the Late Isaac Friedlander," *New York Times*, July 20, 1878; Alfred Bannister, "California and Her Wheat Culture," *Overland Monthly and Out West Magazine* n.s., 12 (July 1888): 65–70; Paul W. Gates, "Public Land Disposal in California," *Agricultural History* 49, no. 1 (1975): 158–178; Book

Club of California, *Breadbasket of the World: California's Great Wheat-Growing Era, 1860–1890* (San Francisco: Book Club of California, 1984).

18. Richard B. Rice, William A. Bullough, and Richard J. Orsi, *The Elusive Eden: A New History of California* (New York: McGraw-Hill, 2002), 223–229; Donald J. Pisani, "Land Monopoly in Nineteenth-Century California," *Agricultural History* 65, no. 4 (Autumn 1991): 15–37; Pisani, *From the Family Farm to Agribusiness*; Donald Pisani, "The Squatter and Natural Law in Nineteenth-Century America," *Agricultural History* 81, no. 4 (2007): 443–463.

19. Rice, Bullough, and Orsi, *The Elusive Eden*, 224–226; Pisani, "The Squatter and Natural Law in Nineteenth-Century America."

20. Menefee and Dodge, *History of Tulare and Kings Counties, California*, 111–112.

21. Benjamin Cooper Wright, *San Francisco's Ocean Trade, Past and Future: A Story of the Deep Water Service of San Francisco, 1848 to 1911* (San Francisco: Carlisle & Co., 1911); Rodman W. Paul, "The Wheat Trade between California and the United Kingdom," *Mississippi Valley Historical Review* 45, no. 3 (December 1, 1958): 391–412; D. Morgan, *Merchants of Grain* (New York: Viking Press, 1979), 78.

22. A. Basil Lubbock, *Round the Horn before the Mast* (London: John Murray, 1902); George Broomhall and John Hubback, *Corn Trade Memories Recent and Remote* (Liverpool: Northern Publishing Co., 1930); R. Perren, "Structural Change and Market Growth in the Food Industry: Flour Milling in Britain, Europe, and America, 1850–1914," *Economic History Review* 43, no. 3 (August 1990): 420–437.

23. J. H. Clapham, *An Economic History of Modern Britain*, vol. 2, *Free Trade and Steel, 1850–1886* (Cambridge: Cambridge University Press, 1967); D. J. Oddy, "Food in Nineteenth Century England: Nutrition in the First Urban Society," *Proceedings of the Nutrition Society* 29, no. 1 (1970): 150–157; Roger Scola, Alan Armstrong, and Pauline Scola, *Feeding the Victorian City: The Food Supply of Manchester, 1770–1870* (Manchester: Manchester University Press, 1992); Ian Douglas, Rob Hodgson, and Nigel Lawson, "Industry, Environment and Health through 200 Years in Manchester," *Ecological Economics* 41 (2002): 235–255; Harold L. Platt, "'Clever Microbes': Bacteriology and Sanitary Technology in Manchester and Chicago during the Progressive Age," *Osiris* 19 (2004): 149–166; Chris Otter, "The British Nutrition Transition and Its Histories," *History Compass* 10, no. 11 (2012): 812–825.

24. Latta, *Uncle Jeff's Story*, 23.

25. "1932 Runoff," Box 99, Folder 6, Latta Collection; *Tulare Daily Register*, January 29, 1889.

26. *Tulare County Times*, January 19, 1899; *Fresno Republican*, December 1, 1929.

27. *Tulare County Times*, June 4, 1898; *Hanford Sentinel*, August 4, 1898; *Fresno Bee*, March 2, 1932.

28. Joshua Yeager, "Tulare County's Ag Industry Grew, But Not as Fast as Fresno's Exploding Almond Crop," *Visalia Times Delta*, October 8, 2019, https://www.visalia timesdelta.com/story/news/2019/10/08/tulare-countys-7-2-b-ag-economy-no-3

-nation/3908035002/; "Kings County Department of Agriculture 2018 Crop Report," https://www.countyofkings.com/home/showdocument?id=20432, accessed December 8, 2020; "The Tachi Yokuts: The Story of the Santa Rosa Rancheria," https://www .youtube.com/watch?v=2rk90s-xT9w, accessed June 11, 2020; "Yokuts Indians of the Central California Valley," ABC News 30, Fresno, CA, August 1, 2009, https://www .youtube.com/watch?v=qjiaXmsRdUM.

29. "'Phantom' Tulare Comes Back to Life," *Los Angeles Times*, February 13, 1997; "Yokuts Indians of the Central California Valley"; M. Kat Anderson, *Tending the Wild: Native American Knowledge and the Management of California's Natural Resources* (Berkeley: University of California Press, 2005), xv, xvii, 8–9; Jane Pleasant, "The Paradox of Plows and Productivity: An Agronomic Comparison of Cereal Grain Production under Iroquois Hoe Culture and European Plow Culture in the Seventeenth and Eighteenth Centuries," *Agricultural History* 85, no. 4 (January 1, 2011): 460–492.

30. For an example of community-engaged scholar-activism, see Paul Routledge and Kate Driscoll Derickson, "Situated Solidarities and the Practice of Scholar-Activism," *Environment and Planning D: Society and Space* 33 (2015): 391–407.

2

Modern Food as Status

A BIOGRAPHY OF MODERN BRITISH BREAD

David Fouser

White bread was perhaps the single most important material object in the lives of Victorian Britons, more so than the bricks of their homes, the fabric on their bodies, the iron that made their tools, or the coal that drove their machines. Nineteenth-century Britain had a ravenous appetite for white bread, but this appetite went largely unsatisfied, for while white bread was the preferred fare, brown bread was far more common. This hunger, long unsated and perhaps insatiable, lay at the heart of British society. So much of Britain's material and cultural life revolved around it that bread was spoken of almost reverentially. The "daily bread" was the "first necessity," "one of the indispensable conditions of life," and, most commonly, "The Staff of Life." "There's nothing like bread," a London street seller told Henry Mayhew in 1850. "It's not all poor people can get meat, but they must get bread."[1]

Bread, like any food, is a package of calories for nourishing bodies that is rooted in environments and enmeshed in culture. In my household, "bread" is an essential item on the weekly shopping list, a staple of breakfast and lunch, but—with the much more varied diet available in twenty-first-century America—often absent from our main family meal, the evening dinner. For Victorian Britons, who could eat a pound or more per day, the word "bread" meant far more than the object it nominally signified. For them, bread sat at the intersection of work and life, production and consumption, and signified material life broadly. Invoking God's command to Adam in the Book of Genesis, people of all ranks spoke of the need to "earn their bread," even those whose wealth came from capital and not, strictly speaking, "the sweat of their brow." "Bread" meant food overall, as Britons gave thanks for their "daily bread" before meals, but it also meant clothing, shelter, and other quotidian necessities. Working-class poetry and song lamented the rags they wore, the crumbling homes they inhabited, and the cold and dark nights they suffered through, all of which could be articulated in a simple cry for "bread."[2]

Just as bread the word signified more than bread the object, the object itself was not a homogenous category. Intertwined in the word's meaning and in its material reality was a core binary: bread was either white or brown. Each signified opposing meanings. White bread was the food of prosperity and freedom, the clearest indicator of one's independence. Working men might bring their children a penny loaf of white bread after a particularly good stretch of work; others made sure to bring white bread with them for their meals while working, even if they ate otherwise at home. Londoners deemed it "derogatory . . . and a sign of poverty," to not eat white bread. Brown bread, in contrast, was the food of poverty, dearth, unfreedom, and, above all, dependence. Brown bread was the fare of discipline and subordination, the food of the workhouse and the prison. It was not simply bread that Victorian Britons wanted, then, but *white* bread.[3]

Wrapped in meanings though it was, the material, bodily dimensions of bread were ever present. They appear on the pages of rural families' diaries, where they recorded harvest failures and years of dearth—shortages of bread, above all—alongside births, marriages, and deaths as the most fundamental moments in their lives. They were there when unemployed men walked miles in search of a day's bread, or when street-sellers and mudlarks (children who combed the muddy banks for the Thames for items of value) walked just as far in order to eke out enough for a dinner of a few slices and a "penn'orth" (penny's worth) of drippings. They were there too when people begged a loaf from a neighbor, when a poor family made a bit of bread soaked in hot water serve as a poor imitation of soup, when children whimpered for the crumbs their parents could not provide, when families chose between coal or bread, and when the most desperate sold their meager possessions and took the money directly to the baker.[4]

Despite an incredible hunger for white bread, most Britons in the early nineteenth century lived on brown bread, as they had for centuries.[5] This began to change in the second half of the century. At that point, the social and environmental relationships that we call "bread" were transformed by free-market capitalism and global settler colonialism. White bread, essentially a medieval luxury, became a modern staple. And, as with so many other objects from fine fabric to suburban houses, the "modern" was the satisfaction of "medieval" desires.

The rise of this new modern loaf gives us a view of how modern foods were made. It is a story of hunger, how it was satisfied, and with what consequences. It begins deep in bread's history, when distinctions between white and brown bread became meaningful. This color line, so visible, so pervasive, was a battleground for bread. Jeffrey Pilcher (chapter 4 in this book) discovers something similar in his study of global beer styles, finding the pale lightness of pilsner one of its main points of appeal. David Singerman (chapter 11) sees the battle over color lines in sugar imports as an axis along which to understand perceptions of better and worse. It is visible here too, in the daily bread of Victorian Britain.

The backstory of bread's color begins with the ancient Greeks and Romans, who saw significance in the distinction between white and brown bread. Those distinctions persisted through the Middle Ages and were institutionalized in England in the thirteenth century through laws called the Assize of Bread, which regulated bread production and sales. The Assize of Bread was a central element of the medieval "moral economy," a set of customs, traditions, and statutes that articulated the relationship between the production and consumption of food, and therefore between humans and their environments, and that typically did so without defaulting to markets and commerce. The "moral" component prioritized community integrity and stability, and held that the connections between production and consumption should be as direct as possible. Ideally, they would be contained within a single community, a single manor, or even a single household, so that people worked the land to grow grain and then baked and ate their own bread from that grain. Scholars have called agriculture practiced in this way "closed-circuit" or "closed-loop" because communities were largely self-reliant; its products might be marketed today as "sustainable" or at least "locally grown."[6]

In the moral economy of late medieval and early modern England, any individual or entity that stood between the production and consumption of food was a morally suspect middleman at best, a criminal at worst. The Assize of Bread institutionalized this moral framework, at least where it was necessary for people to buy bread. It did so by setting prices for bread based on the price of wheat. Since wheat prices were essentially reflections of local weather and agriculture, it removed—or at least displaced— direct human agency from the question of bread prices. God's wrath might bring drought or blight, leading to dearth, high prices, and empty bellies, but that was largely out of human hands.[7]

In regulating the price of bread, the Assize specified qualities. Early versions of the law defined varieties from "Wastel," a fine, white bread with the highest price, to "Cocket," "French," "Ranger," and "Bread Treet," each with specifications as to quality, ingredients, methods, and price. In this way the Assize of Bread was about consumption too, for it determined what bread might be consumed and to some extent who might consume what. It effectively limited the quantity of white bread that might be baked and thereby invested white bread with status, prestige, and privilege. Brown bread was simultaneously marked as the bread of want, poverty, and dependence. Over succeeding centuries the many categories of bread named in the Assize collapsed, becoming essentially two, "Wheaten" for white bread and "Household" for brown, while the difference in status between the two remained. The Assize eroded over the centuries before being repealed completely in London in 1815 and across the kingdom in 1836, but it left a lasting imprint on British culture and diet: white bread was essentially always and everywhere both in high demand and in short supply.[8]

This doesn't mean that bread's long life from thirteenth-century England to nineteenth-century Britain was uneventful: over those centuries, grain and bread became much more commercialized, and as a result the kingdom's bread became at least a bit whiter overall. Merchants and millers inserted themselves into networks of grain production, distribution, and consumption, and by the late eighteenth century nearly all agricultural production was directed toward markets rather than to their communities, as the moral economy originally imagined. Most food supplies remained local until around 1850, but British merchants did begin to import grain regularly to feed growing cities. Imports reached as much as a quarter of the total supply by the middle of the nineteenth century, a portion that only increased thereafter. Some of it came from California's new wheat fields, as Thomas Finger describes in chapter 1.[9]

Bread became somewhat whiter overall as it became more commercial. Profit-seeking farmers grew wheat because it brought better prices, millers ground grain not according to ancient feudal traditions but rather for their own profits, and bakers catered to demanding consumers who demanded the whitest bread they could get.[10]

Still, brown bread remained the fare for most Britons because wheat and bread have material limits. Up to the middle of the nineteenth century these limits lay in Britain's climate and the technology available to British millers and bakers. In the damp, temperate climate of the British Isles the wheat that grew best was typically white in color, physically soft, and relatively moist. It was "weak" in baking terms, with relatively low levels of gluten. Gluten is the main protein created when flour and water are kneaded together to make dough, and as such it is the main ingredient in the gooey matrix of flour and water that, when charged with gas in some way (usually by yeast), gives leavened bread its characteristically porous structure and light texture. Low-gluten flour necessarily meant dough that could not capture as much gas and therefore resulted in bread that was fairly dense in texture. Or, at the very least, it took high-quality flour and considerable skill on the part of bakers to bake truly white, lightly textured, and well-risen bread. It was in this way that climate influenced bread's character.[11]

As with climate, milling was also a factor in bread's identity. The equipment of British flour milling in the mid-nineteenth century retained deep roots in the medieval period, contributing directly to the dominance of brown bread. While the business of milling had changed considerably since the thirteenth century, their roughly four-foot-diameter millstones would have been recognizable to medieval millers, and perhaps even to their ancient Roman forebears.[12] Once ground, the miller "bolted" the resulting meal, sifting it with progressively finer-meshed sieves. These removed the bran and some of the germ. Following its initial grinding and bolting, successive sieves removed more and more of the "pollards," the remaining bits that were anything but the starchy endosperm of the wheat. That starchy endosperm, when granulated, became flour.

This whole process had a tremendous impact on the nature of the flour, which was in turn the central element determining the color of the resulting bread. And here lay a dilemma: the more thoroughly the meal was bolted, the whiter the resulting flour but the smaller the amount produced. When they were governed by medieval customs, millers had little opportunity to maximize their profits by making more white flour. The erosion of those customs did allow millers to produce more whiter flour but they were still limited by their equipment, which simply could not produce as much white flour as the kingdom's collective appetite craved. This remained the case into the middle of the nineteenth century, when most bread remained fairly brown in color and relatively dense in its structure.[13]

Given these limitations, bakers did what they could to meet that demand—a demand that was so high, in fact, that bakers almost universally adulterated their flour with alum, a bleaching agent. It was illegal on the grounds that it was fraudulent and poisonous, but every account of baking through at least the 1860s acknowledged its universal use all the same. Not surprisingly, when *The Lancet* investigated food purity and adulteration in the 1850s, every single loaf of bread they tested had alum in it. Indeed, fears of impurity may have actually helped the popularity of white bread over brown, even as it simultaneously encouraged its adulteration so that it would appear white. While alum was universally used, other, more sinister adulterants lurked on the pages of sensational pamphlets; there were occasional claims of sawdust and chalk mixed into the dough, and even of ground-up bones. Bakers noted that it would be more difficult, and perhaps even more expensive, to make bread with sawdust in it—never mind some kind of bone meal or chalk—but in the laissez-faire world of Victorian Britain there were no assurances of the purity of food other than one's senses. The best one might do is buy white bread if possible, which at least *looked* pure.[14]

Such is the backstory that brings us to a period of relatively rapid change. British bread at the midpoint of the nineteenth century resembled its medieval ancestors in fundamental ways: its origins were essentially local, its methods and meanings rooted deeply in the past. And yet, by the end of the century, British bread had been transformed. Between about 1850 and 1900 its origins became global, its methods of production were updated, and the meanings associated with white and brown bread shifted in significant ways. Bread itself became "better," whiter, and more highly risen, a fact that all agreed on by the early twentieth century. British bread became "modern," at least to the extent that ancient hungers were satisfied: white bread went from a medieval luxury to a modern staple.

The causal element in the transformation of Britain's bread was the globalization of its wheat, brought about by free-market capitalism and global imperialism. It was this change that produced technical consequences and ultimately changed Britons' daily bread. When we look for the origins of white bread, we look to the second half of the nineteenth century.

As an expression of the relationships between production and consumption, bread is a relationship between environments and humans. British bread until about 1850 reflected British environments: the damp, temperate climate produced white wheat that was soft, weak, and moist; British milling was well suited to process such wheat and did so in the context of the Assize of Bread. Within those conditions white bread was privileged over brown, ultimately leading to a society in which truly white bread was limited and the bulk of bread actually consumed was fairly dense in texture and adulterated with alum to look white.[15]

Long and broad processes like globalization do not often have clear inflection points, but in the case of the globalization of Britain's bread it does: 1846. In that year, free trade became law in Britain with the repeal of the Corn Laws. The Corn Laws were duties on imported grain intended specifically to keep the price of grain high and thereby prop up rental incomes for landlords. Their repeal in 1846 was a central component of nineteenth-century British liberalism, the philosophical bedrock of free-market capitalism. It was at this moment that free trade became the enacted policy of the United Kingdom. Free trade of course mattered little unless there were goods to trade, and thus the expansion of European and American empires provided an essential ingredient in the transformation of British bread. Global imperialism brought far greater amounts of wheat into Britain.[16]

In North America, Argentina, and Australia, settlers expropriated and in some cases exterminated Indigenous inhabitants. In their place, the settlers seized land and brought those lands under the plow. In the United States, Canada, Argentina, and Australia they cleared forests, drained swamps, plowed grasslands, and irrigated deserts, transforming diverse environments into wheat-producing farmland. Here again we can look back to chapter 1's tale of Central California's dried-up lakebeds to illustrate this process from the perspective of California, looking outward as a region of agricultural export. Wheat was hardly the region's only crop, but it was among the most popular choices because it stored and traveled better than most other commodities and because it was value-dense, bringing relatively high prices for its weight. In the Old World the process was somewhat different. In India, southern Russia, and Ukraine, local populations were largely coerced through legal mechanisms into adopting wheat agriculture, but there too the result was in crucial respects the same: environments around the world were transformed, becoming parts of a global wheat ecology.[17]

This chapter provides a view of the transformation from Britain. There, free-market capitalism and global imperialism combined with the rapid development of fossil fuel transportation networks to produce a dramatic increase in British wheat imports in the later 1800s and a corresponding fall in prices. Imports were typically cheaper than homegrown wheat, and as imports grew in the second half of the century the overall price dropped substantially. The transformation of British bread provides an object lesson in how specific government philosophies (in this case, market liberalism) lead to specific trade policies (here, free trade) which in turn generated changes in land use patterns abroad and finally resulted in utterly transformed diets. In 1846 Britain imported about 25 percent of the wheat that it consumed; by 1870 the figure was just over 50 percent, and by the outbreak of World War I in 1914 it was a staggering 80 percent. In absolute terms, Britain imported five times more wheat in 1900 than it had in 1850 and the price had fallen by roughly half. Britain was by far the largest importer of grain in the world, and one merchant guessed that more than half of all wheat imports globally went into Britain during this period.[18]

The globalization of wheat transformed British bread because wheat from the rest of the world was quite different from wheat grown in Britain. All agricultural products vary considerably. In particular, foreign wheat was typically better suited to producing whiter, more highly risen, lightly textured bread. Wheat can be red or white, hard or soft, strong or weak, moist or dry. It can also fall anywhere on a broad, multidimensional spectrum of flavor. Much of this variation is linked to the environments in which crops are grown, and the environments that comprised the global wheat ecology in the late nineteenth century were nearly all drier and less temperate than Britain. The North American plains, Argentine pampas, Australian wheatlands, and Ukrainian black soil regions were all drier than Britain, had hotter summers, and in some cases colder winters. India's wheat-growing environments were somewhat more varied in their moisture, but they too were universally hotter than Britain. As a result these environments produced wheat that was physically harder, drier, and stronger in baking terms. It was also often red wheat, with darker husks and bran.[19]

These characteristics, particularly when combined with lower prices, made imported wheat attractive to British bakers, at least in theory. They wanted stronger wheats, which had significantly more gluten than the weak British wheats they were accustomed to. With more gluten the dough could capture more gas in fermentation and bake into taller, more highly risen and more lightly textured loaves. Bakers even found that with strong flour and new forms of industrial yeast they could cut the time it took to ferment and bake bread in half, from twenty-four hours or more to just twelve. British bakers were thus enthusiastic proponents of imported wheat for it offered them the opportunity to bake the loaves that their customers demanded more quickly.[20]

Even if the rise of white bread was relatively swift, it wasn't necessarily easy. If foreign wheat offered bakers the possibility of better bread, for instance, it brought particular challenges for millers. Each farmer's particular soil and techniques varied. They rotated their crops in any number of ways, employed more or less labor at different points in the agricultural cycle, used or discarded various tools or machines, and battled their particular weeds and pests more or less successfully. Even the same land, worked by the same people, could produce quite different products from year to year as the weather varied. All of these factors affected how millers actually handled each batch of grain. For example, millers frequently had to clean wheat before it was milled, but just how it had to be cleaned could vary tremendously: did they need to remove dust and small parti-cles of dirt and mud? Were there bits of straw, stones, or other, more exotic "additives" or "impurities," such as pieces of metal or foreign seeds? All of these had to be removed, but all required different treatments and all var-ied according to the particular source of wheat. Russian wheat often had linseed because, as far as British millers could tell, Russian farmers often rotated wheat onto lands after crops of linseed. Indian wheat was often unusually dirty; it seems that British merchants in Calcutta, Bombay, and Karachi refused to pay premium prices for Indian wheat, and Indian cul-tivators then deliberately neglected cleaning their grain since they would get no benefit from it and might have even resorted to adding stones and dirt to increase the volume.[21]

Imported wheat was so different from British-grown wheat that even in the best of circumstances it was difficult to handle, and could even be explosive. Grinding grain from California or southern Russia on millstones was a far different experience from grinding grain from Essex or Norfolk. Most imported wheat was darker in color (again, red wheat as opposed to the white wheat most British farmers grew), physically harder, and signif-icantly drier than British wheat. Millers discovered that imported wheat could catch fire as the hard, dry kernels ignited in the intense friction between the millstones, and in the highly combustible atmosphere of a flour mill this could be disastrous. Even when millers avoided catastro-phe, the grains themselves were stubborn materials to handle. "The bran and offals are so hard and brittle, and adhere so closely to the kernel," one miller complained about North American wheat, "that when ground in the ordinary way it is practically impossible to keep them from getting ground into the flour." This darkened the flour, which of course looked worse, was more difficult to bake with and was consequently unpopular among bakers. That, along with fire hazards, made for an uneven adjust-ment to new wheats. All of the adjustments millers and bakers made sug-gest that white bread was neither inevitable nor straightforward.[22]

In fact, Britons developed two main solutions to the problems of imported wheat so that its nearly universal uptake, in retrospect, could look inevitable. In the first place, millers had to learn far more about global wheat supplies than ever before. Some millers before midcentury had dealt with imports—they were a quarter of Britain's wheat supply, after all—but millers' and bakers' commentary on the subject emphasized the unpredict-able nature of foreign grain. And the great majority of mills prior to that point were located in British grain-growing districts, often surrounded by the very wheat they ground.

Millers had long-standing relationships with their sources and might even buy grain sight unseen if they knew the farmer or merchant well enough. But by the 1870s, producing quality flour for Britain's bread required a far more cosmopolitan approach. For one, British millers began to carry out systematic studies of global wheat, assessing the milling and baking qualities of each variety. For another, they became experts in the hundred or so varieties of wheat that came to dominate the growing mar-kets at London and Liverpool. This gave them the expertise to choose their supplies wisely and in most cases to choose mixtures of different kinds of wheat from around the world. A skilled miller could manage the moisture content, strength, and flavor of the flour they produced, even when the particular supplies fluctuated from year to year—one year Russia might have a bumper crop, thus dropping its price, while the next year that might be true of American wheat.[23]

The second solution to the problems of foreign wheat was industrial: pairs of steel rollers replaced the old millstones. Roller milling was not indigenous to Britain. The earliest use of steel rollers was in the Hungar-ian "Walzmühle" in Budapest in 1839. American millers in Minneapolis were the first to develop integrated and automated "milling systems" that incorporated rollers in the 1860s. By the 1870s British millers feared Amer-ican flour imports and saw the potential in global wheat supplies enough to begin to adopt them en masse. By the 1880s the number of rural "coun-try" mills, grinding the grain of British fields, was dropping precipitously because, thanks to Empire-assisted globalization, British millers found it far more useful to locate in port cities that granted them ready access to the world's wheat supplies. By the end of the nineteenth century the great bulk of British milling was found in just five ports: London, Liverpool, Glasgow, Hull, and Bristol.[24]

Again, this was a technological change encouraged by geographical differences. Steel rollers changed the game for British millers by permit-ting more variability and precision. They could be smooth or fluted in different patterns and could rotate at variable speeds. Most importantly, the space between the two rollers could be adjusted minutely, allowing a much different approach, known as "gradual reduction." The new "gradual reduction" or "high grinding" method allowed them to slowly and carefully remove offal and to produce a far greater proportion of genuinely white flour than was possible with millstones. While traditionally millers had been able to produce a small proportion of very fine flour or larger proportions of "seconds" flour, roller mills could produce "short patents" of 30 percent extraordinarily white flour or "long patents" of 60 percent white flour.[25]

With the creation of a global wheat supply and roller flour milling, the central ingredient of British bread changed dramatically in the space of a few decades. British bread rapidly became whiter, lighter in texture, and more highly risen. Although a preference for white bread had been almost universal prior to the second half of the nineteenth century, the demand had always far outstripped the supply. By the early twentieth century, all agreed that bread was "better"—and by that they meant whiter. Millers and bakers alike spoke of the "craze for whiteness" in bread and began to use terms like "ultrafine" to describe the highest grades of roller-milled flour. Bakers extolled its virtues, telling anyone who would listen that nothing but the whitest bread would sell, claiming that bread had to be handsome and "white as snow" or their customers would take their business elsewhere. Modern white bread was born.[26]

So profound was the transformation of Britain's bread that the ancient meanings attached to white and brown bread began to shift. For centuries white bread had been privileged over brown, and that privileging was essentially premised on a world in which white flour was comparatively rare or was made such through legislation. But as early as the 1860s British bread was already becoming much whiter than it had been, and by the end of the century British working people were able to afford it more consistently. Workhouses—for a generation defined by their poor diets—stopped insisting on brown bread after a Parliamentary committee admitted that white bread was cheaper and more physiologically efficient than brown.[27]

Ever mindful of class distinctions, a segment of the bourgeoisie responded to the proliferation of white bread among the working classes by adopting brown bread. Once the signifier of want, poverty, and dependence, brown bread became a bourgeois symbol of health, wisdom in consumer choice, and respect for tradition, much as it is today. Conversely, white bread's most intense articulation, Wonder Bread, is today widely regarded as *the* symbol of unsophistication.[28] This flip-flop is a consequence of the changes wrought over the course of the nineteenth century, and it's one that persists to this day: in my own household, we distinguish between "proper, nice bread"—a ciabatta loaf, a sourdough batard, a baguette, something bought fresh that stays on the counter until it's consumed, usually in a day or two—and "regular bread," the industrial, presliced loaves that inhabit the freezer and are typically relegated to sandwiches packed in lunchboxes.

By the turn of the twentieth century, then, British bread had become "modern": materially its origins had become global and its physical nature and characteristics were transformed so that it was whiter overall. Culturally, the meanings invested in white and brown bread and what they alleged to signify about one's place in the social hierarchy were inverted. No longer was brown bread the bread of poverty and dependence, and white bread the bread of privilege, prosperity, and independence. Instead, what was a medieval luxury became a modern staple.

NOTES

1. Terms such as "daily bread" and "Staff of Life" are truly ubiquitous in sources of all kinds from nineteenth-century Britain. George Dodd, *The Food of London: A Sketch of the Chief Varieties, Sources of Supply, Probable Quantities, Modes of Arrival, Processes of Manufacture, Suspected Adulteration, and Machinery of Distribution, of the Food for a Community of Two Millions and a Half* (London: Longmans, Brown, Green and Longmans, 1856), 167; Eliza Acton, *The English Bread-Book for Domestic Use* (London: Longman, Brown, Green, Longmans, & Roberts, 1857), v; Henry Mayhew, *London Labour and the London Poor* (s.n., 1851), 228.

2. See, for example, the writings of Chartist activist William Lovett, *The Life and Struggles of William Lovett: In His Pursuit of Bread, Knowledge and Freedom; With Some Short Account of the Different Associations He Belonged to and of the Opinions He Entertained* (London: Trübner & Co., 1876); Herbert Burrows, *A Song of Freedom* (London: Labour Emancipation League, 1880).

3. *Hungry Forties: Life under the Bread Tax; Descriptive Letters and Other Testimonies from Contemporary Witnesses* (London: T. Fisher Unwin, 1904), 94–95; Lovett, *The Life and Struggles of William Lovett*, 13; Dodd, *The Food of London*, 204; A. S Krausse, *Starving London: The Story of Three Weeks' Sojourn among the Destitute* (London: Remington & Co., 1886), 21, 33; J. R. Widdup, *The Casual Ward System: Its Horrors and*

Atrocities; Being an Account of a Night in the Burnley Casual Ward, Disguised as a Tramp (London: Labour Press Society, 1894), 5, 8, 16.

4. Lovett, *The Life and Struggles of William Lovett*, 13; Mayhew, *London Labour and the London Poor*, 153, 188, 216, 269; Krausse, *Starving London*, 20–22, 32–33; Edwin Waugh, *Home-Life of the Lancashire Factory Folk during the Cotton Famine* (London: Simpkin, Marshall, 1867), 209–213.

5. E. J. T. Collins, "Dietary Change and Cereal Consumption in Britain in the Nineteenth Century," *Agricultural History Review* 23, no. 2 (1975): 97–98; Michael Nelson, "Social-Class Trends in British Diet, 1860–1900," in *Food, Diet, and Economic Change Past and Present*, ed. Catherine Geissler and Derek J. Oddy (Leicester, UK: Leicester University Press, 1993), 104–105.

6. E. P. Thompson, "The Moral Economy of the English Crowd in the Eighteenth Century," *Past & Present*, no. 50 (February 1971): 76–136.

7. James Davis, "Baking for the Common Good: A Reassessment of the Assize of Bread in Medieval England," *Economic History Review* 57, no. 3 (August 1, 2004): 465–502; Alan S. C. Ross, "The Assize of Bread," *Economic History Review* 9, no. 2 (1956): 332–342; Sidney Webb and Beatrice Webb, "The Assize of Bread," *The Economic Journal* 14, no. 54 (June 1, 1904): 196–218.

8. Davis, "Baking for the Common Good," 471; George Read, *A Brief History of the Bread Baking Trade, from the Earliest Period to the Present Time* . . . (London: George Biggs, 1848), 4–5; Thompson, "The Moral Economy of the English Crowd in the Eighteenth Century."

9. C. R. Fay, "The Miller and the Baker: A Note on Commercial Transition, 1770–1837," *Cambridge Historical Journal* 1, no. 1 (1923): 85–91; Christian Petersen, *Bread and the British Economy, c. 1770–1870*, ed. Andrew Jenkins (Aldershot, UK: Scolar Press, 1995); Karl Gunnar Persson, *Grain Markets in Europe, 1500–1900: Integration and Deregulation* (Cambridge: Cambridge University Press, 1999); Roger Scola, *Feeding the Victorian City: The Food Supply of Manchester, 1770–1870* (Manchester, UK: Manchester University Press, 1992).

10. Petersen, *Bread and the British Economy*, 15–20; Collins, "Dietary Change and Cereal Consumption in Britain in the Nineteenth Century," 97–98.

11. This is a central argument of my dissertation. See David Fouser, "The Global Staff of Life: Wheat, Flour, and Bread in Britain, 1846–1914" (PhD diss., University of California, Irvine, 2016), and the sources therein.

12. Adam Lucas, *Wind, Water, Work: Ancient and Medieval Milling Technology* (Leiden: Brill, 2006), 20–22; Glyn Jones, *The Millers: A Story of Technological Endeavour and Industrial Success, 1870–2001* (Lancaster, UK: Carnegie Publishing, 2001), 22–23; "History of the Decortication of Wheat," *The Miller*, October 2, 1876.

13. References to "seconds" flour are ubiquitous, but see for example Acton, *The English Bread-Book for Domestic Use*, 87–89.

14. E. J. T. Collins, "Food Adulteration and Food Safety in Britain in the Nineteenth and Early Twentieth Centuries," *Food Policy* 18 (1993): 95–109.

15. Petersen goes so far as to suggest that perhaps alum made possible the consumption of otherwise unmarketable bread. See Petersen, *Bread and the British Economy*, 119–120.

16. Giovanni Federico, *Feeding the World: An Economic History of Agriculture, 1800–2000* (Princeton, NJ: Princeton University Press, 2005); Ronald Findlay and Kevin H. O'Rourke, *Power and Plenty: Trade, War, and the World Economy in the Second Millennium* (Princeton, NJ: Princeton University Press, 2007).

17. The literature on settler colonialism is vast, but see for example William Cronon, *Nature's Metropolis: Chicago and the Great West* (New York: W. W. Norton, 1991); Jeremy Adelman, *Frontier Development: Land, Labour, and Capital on the Wheatlands of Argentina and Canada, 1890–1914* (Oxford: Oxford University Press, 1994); and James Belich, *Replenishing the Earth: The Settler Revolution and the Rise of the Anglo-world* (Oxford: Oxford University Press, 2011) for good accounts.

18. B. R. Mitchell, *British Historical Statistics* (Cambridge: Cambridge University Press, 1988), 221–226.

19. Fouser, "The Global Staff of Life," chaps. 2 and 4.

20. Fouser, "The Global Staff of Life," chaps. 2–4. For one of many contemporary sources dealing with this topic, see William Jago, *A Text-Book of the Science and Art of Bread-Making: Including the Chemistry and Analytic and Practical Testing of Wheat, Flour, and Other Materials Employed in Baking* (London: Simpkin, Marshall, Hamilton, Kent, & Co., 1895).

21. Fouser, "The Global Staff of Life," chap. 2; Edward Bradfield, *Wheat and the Flour Mill: A Handbook for Practical Flour Millers* (Liverpool: Northern Publishing Co., 1920), 25, 38–39, 68; Jago, *A Text-Book of the Science and Art of Bread-Making*, 288–295, 299–300; William R. Voller, *Modern Flour Milling*, 3rd ed. (New York: D. Van Nostrand, 1897), 11; Arthur Barker, *The British Corn Trade, from Earliest Times to the Present* (London: Sir Isaac Pitman & Sons, 1920), 83; J. H. Chatterton, *Eleventh Annual Report of the National Association of British and Irish Millers* (London: George Berridge & Co., 1889), 20. For the first full survey of Indian wheat, see John Forbes Watson, *Report on Indian Wheat* (London: HMSO, 1879).

22. Fouser, "The Global Staff of Life," chap. 2; Jones, *The Millers*, 11; "Scientific and Practical Milling, No. IV," *The Miller*, February 12, 1876.

23. Fouser, "The Global Staff of Life," chap. 4. For contemporary commentary on these topics, see John White, *A Treatise on the Art of Baking* (Edinburgh: Printed by Anderson & Bryce, 1828), 139; George Broomhall and John Hubback, *Corn Trade Memories, Recent and Remote* (Liverpool: Northern Publishing Co., 1930), 7–43. On the technical development of the milling industry, see the trade publications that emerged in the 1870s, such as the Liverpool-based journal *The Miller*.

24. Fouser, "The Global Staff of Life," chap. 2; J. H. Chatterton, *Ninth Annual Report of the National Association of British and Irish Millers* (London: George Berridge & Co., 1887), 114; Hugh J. Sanderson, *Seventeenth Annual Report of the Transactions of the National Association of British and Irish Millers* (London: George Berridge & Co., 1897),

100–101; Jones, *The Millers*, 22–35; Richard Perren, "Structural Change and Market Growth in the Food Industry: Flour Milling in Britain, Europe, and America, 1850–1914," *Economic History Review*, n.s., 43, no. 3 (August 1990): 420–437.

25. The technical details of milling are found in the substantial contemporary literature, as noted above; the specific example here is from John Kirkland, *The Modern Baker, Confectioner, and Caterer: A Practical and Scientific Work for the Baking and Allied Trades* (London: Gresham, 1913), 52–55.

26. Discussions of the demand for white bread were common; two examples, from consumers and from producers, are *Hungry Forties*, 28–29, 69–70, 93–94, 138–139, and Henry Simon, "On the Latest Development of Roller Flour Milling," in *Proceedings of the Institution of Mechanical Engineers* (London: Institution of Mechanical Engineers, 1889), 180, 190.

27. Edward Smith, "Dietaries for the Inmates of Workhouses," Command Papers no. 3660 (London: House of Commons, 1866), 36. For a broader discussion of workhouse diets, see Valerie J. Johnston, *Diet in Workhouses and Prisons, 1835–1895* (New York: Garland, 1985).

28. Aaron Bobrow-Strain, *White Bread: A Social History of the Store-Bought Loaf* (Boston: Beacon Press, 2012).

3

Modern Food as Colonized Food

OLD IS BAD, NEW IS AMERICAN

PHILIPPINE FOOD CONSUMPTION AND PRODUCTION DURING AMERICAN EMPIRE IN THE EARLY 1900s

René Alexander D. Orquiza Jr.

In the aftermath of the Philippine-American War, in the first decade of the twentieth century, American imperialists turned to food to justify a new policy of empire and investment in the archipelago. They labeled Filipino cuisine uncivilized, claimed that it needed the corrective hand of Western domestic science, and urged investment in newly industrializing Filipino agriculture. American authors of children's books, travel guides, scientific pamphlets, and business manuals focused on food as they appealed to an American public weary from nearly five years of fighting that had started with the Spanish-American War in 1898 and closed with the suppression of Philippine independence in 1902. Food became a way of expressing the supposed racial inferiority of Filipinos and the dawn of a new era of extracontinental Manifest Destiny. The Philippines was also a booming arena for wildcat speculation in America's first Asian colony. Rather than enhance American knowledge of food in the Philippines, accounts of Filipino food over the next half century contributed to ignorance that, in many cases, persists with modern Filipino food as the unknown cuisine of Southeast Asia in the United States today. Perceptions of food shaped many of the American racial attitudes and misconceptions of Filipinos. Here, modern food involved imagining a new place—an American fantasy of the colonized Philippines.

I have lived a particular personal experience of this fantasy. Growing up in San Jose as a Filipino American in the 1990s, I often was asked, "Why aren't there more Filipino restaurants when there are so many Filipinos in the Bay Area?" I heard it from friends new to the region, budding foodies eating their way through San Francisco, and fellow locals who had no clue about the mom-and-pop carinderias in their towns. When I worked in restaurants after college I heard it from fellow cooks and chefs who had moved to San Francisco specifically because it was an American culinary jewel. I then moved to Baltimore, a city with even fewer Filipinos, and the question morphed into this: why didn't Americans know about Filipino cuisine?

The answers lie in the history of how Americans belittled, racialized, and denigrated Filipino cuisine in an imperial relationship beginning in 1898. There are few Filipino restaurants and little knowledge of Filipino cuisine because the momentum of colonization pushed American ways to the Philippines without allowing Filipino food to travel back to the United States. It's an unsavory story that recalls the dehumanizing racism and colonial exploitation of American empire that most would rather forget. This imperial rule elevated American cuisine in the Philippines but never acknowledged the legitimacy of Filipino cuisine in the United States. Instead, Americans inextricably linked Filipino food with racism toward Filipinos and used it to justify the American empire's exploitation of Philippine workers and natural resources. They prioritized American imperial objectives over the long-term development of Philippine economic and ecological concerns, creating Philippine foodways that flowed in only one direction—from the United States to the Philippines. These racial associations were so powerful that even as Filipino Americans emerged as the second-largest Asian American population, many Filipino Americans were still reluctant to share their foods with Americans. The result is a cuisine that to this day struggles to gain acceptance and free itself from the imperial connotations given by Americans a century ago.

At the conclusion of the Philippine-American War in 1902, many Americans turned to changing Filipino cuisine as a cheerful, tangible way of justifying the war, one of the most sordid episodes in American history. The three-year conflict resulted in the deaths of over 6,000 Americans and 1.2 million Filipinos, the majority of whom were civilians. The war determined that, after 300 years of Spanish colonial rule, the Philippines would be governed not by Filipinos who had been fighting for independence but by Americans who had purchased the Philippines from Spain for $20 million. Quelling that independence movement required three years of brutal violence and atrocities. The war was the first time the US military used waterboarding, euphemistically called the "water cure," as a method of torturing enemy combatants. American newspapers repeatedly featured images of dead Filipino soldiers piled high alongside roads and buried deep in mass graves as proof that the conflict was going Uncle Sam's way. A small but vocal anti-imperialist minority opposed the war based on a combination of humanitarian and economic grounds. They appealed to American reason by citing the routine torture of Filipinos and the likelihood that Filipinos would compete with Americans for jobs. Mark Twain criticized the war by stating that the stars on the American flag ought to be replaced with skulls and crossbones and the white stripes painted black. While anti-imperialists viewed the conflict with alarm, most Americans considered the war a necessary step in the ongoing mission of Manifest Destiny. Moreover, they believed that successfully remaking the Philippines in an American image would prove that the United States had surpassed European powers in the game of empire.

American proponents of empire began presenting Filipino cuisine as inferior to American cuisine, specifically the recipes and procedures of American domestic science. These authors presented American food by creating racialized associations of food, questioning Filipino cuisine's nutrition, and equating Philippine etiquette with barbarity. Listen to Caroline Shunk, the wife of an American soldier in Manila, who described Filipino food as unappetizing in 1914 because Filipinos "do not cook their food, but eat the flesh of animals and fishes raw, tearing it with their sharpened teeth."[1] Or an American who, in 1904, sardonically described *kamayan*, or the native custom of eating with hands instead of utensils, as a sign of an uncivilized society. "Table manners are unknown and well, hardly required, as every one [*sic*] helps himself, securing the food with his (or her) natural and five-pronged fork—the hand."[2] Another soldier advised American travelers in 1898 not to think too hard about what they were eating because "the less one sees of native cooking, in situ, the greater will one's appetite be."[3] The soldier went on to mock the banquets of Filipino elites as poor facsimiles of Western feasts because "every dinner was practically the same."[4] Defining Filipino cuisine as a single set of characteristics ignored the diversity across its 7,000 islands. Yet this did not stop many Americans from dismissing all Filipino cuisine as terrible.

Criticizing Filipino cuisine was a part of the effort to present Filipinos as colonial subjects who desperately needed American assistance. Help from the United States would come primarily by redefining Philippine food consumption. Improving Filipino food from its supposedly unrefined and unsanitary state would require an American-controlled government, American-run businesses and investments, and an infusion of American teachers to transform how Filipinos thought about cooking dishes. *Our Islands and Their People*, an 1899 publication that championed American control of the Spanish-American War colonies, described an average restaurant in Manila as a haven for shady Filipinos. "It is a rare thing to see a pleasant face among these people," it stated before disparaging Filipino people in general. "As a race, they are vindictive and treacherous—just the kind of people that all good Americans desire to keep away from."[5] Another American author wrote in 1908 that the best restaurants in Manila did not hire Filipino cooks because they were "very poor and indolent" and were "not as industrious as either the Japanese or Chinese."[6] Food allowed authors to establish an American twist on Philippine racial inferiority that detached Indigenous food from any sense of modern civilization. This process led American readers at home to interpret the transformation of the Filipino diet as a benevolent act in the larger improvement of Philippine society.

To inspire well-to-do Filipinos to change their eating habits, American authors focused on what they viewed as the positive change in Manila's increasingly international restaurant scene and the quick embrace of American domestic science around the country. *Harper's* magazine boasted in 1916 that American reformers had effectively standardized the hygienic practices at public markets and religious festivals so that barrels of disinfectant were "almost as conspicuous as the statues of the Virgin."[7] An American soldier wrote that early on in the Philippine-American War, one could thankfully find luxury items such as canned oysters, broiled chicken, and wine and champagne from Europe in Manila's finest restaurants.[8] Manila's American expatriate elite embraced food as a sign of success, such as businessman John Lewis recounting in 1900 how lavish meals were staples of his daily life:

I have fallen into the daily routine of Oriental residents. At six thirty I arise, have breakfast, take a long walk and return to the hotel for a bath and a change of clothes. At nine I am at the office where I stay until twelve. Then to the hotel for tiffin a cigar and a nap . . . [at five] we get into a carriage and drive up and down the Luneta listening to the music and watching the people until quarter of seven, when we drive to the taking some light refreshments en route. Theatre commences at seven and lets out at half past nine. . . . At ten we have dinner at the hotel; then, after a cigar and a song or two from some one [*sic*], to bed.[9]

Banquets and imported goods were markers of the rich, yet the adoption of American domestic science affected the Philippine masses. American reformers eagerly championed domestic science texts that endorsed American cuisine and the belief of American racial superiority. A 1922 cookbook, *Good Cooking and Health in the Tropics*, told American readers they had to supervise Filipino domestic servants because of their supposedly inferior mental capacity and their propensity to lie.[10] Other authors presented domestic science as a means of progress. A 1914 article described proper food instruction as a tool for elevating the country's poor by bringing the standards of "a first class hotel or a wealthy private home" into the ubiquitous "huts with no conveniences" around the country.[11] Domestic science texts were a key part of spreading new approaches to food so that by the 1920s, American teachers were advocating that "every library, whether the library of the school or the private library of the teacher, should be supplied with this [domestic science] material."[12] The rapid adoption of American domestic science supposedly proved that Americans could administer and affect the lives of millions in their new colony just as effectively as their European counterparts.

As Filipinos learned American domestic science and American dishes, they never gained control of the production and export of Filipino food items. In contrast, American control of Philippine farming and agricultural science further made food into a tool for exercising American imperial power over Filipinos.

Scientists, businessmen, travel writers, and government bureaucrats all used food to promote American agricultural investment as a way to further control the Philippines. Their descriptions of the Philippines largely removed Filipino people and cuisine to focus instead on the country as the latest opportunity for Gilded Age land speculation. One American investor excitedly boasted in 1898 that Americans could tap the land for inevitable riches. "There is no telling what heights the Philippines might not attain [with] Yankee push and enterprises [and] the American's natural bent for innovation."[13] Others saw the Philippines as a new American frontier, ripe for pioneer settlement reminiscent of the recently concluded Plains Indian Wars. The author in that case, Hamilton Wright, rhapsodized in 1907, "the resources and labor are there in abundance, and the invitation presented to-day is like that of the once unsettled West, when cheap and fertile land thronged with homeseekers."[14] Politicians joined the fray, too. One, Theodore W. Noyes, supported the construction of railroads that would allow Americans in the Philippines to both bring agricultural items to market and to administer colonial government.[15] That vision, too, helped American investors associate the export of food with the benevolent uplift of Filipinos.

To link the Philippine economy to American market control, government scientists also presented Philippine food exports as the basis for the new Philippine economy. Coconuts, sugar, and rice, for example, would no longer be staples of Filipino agriculture for the sake of Filipino consumers but as commodities ready for the global market. The American-led Philippine Bureau of Science helped in this regard (or hurt, as the case may be), publishing pamphlets and brochures that promoted specific cash crops for American investors. *The Philippine Coconut Industry*, a treatise from 1913, encouraged readers to see coconuts not as food but as fuel and the nation's most profitable export as an affordable alternative to fossil fuels such as coal and oil. The treatise labeled the coconut "the world's most important food fruit" not for its edible meat but for its two by-products, copra and coconut oil, which served as alternatives to fossil fuels.[16] It also framed the modernization of Philippine farming as a question of American patriotism: Spain had made a mess out of Philippine agriculture, its author wrote, and now it was up to the United States to turn the tide—and make a healthy profit along the way. Government scientists also highlighted sugar, the country's second-most profitable commodity and one that, as Benjamin Cohen (chapter 6) and David Singerman (chapter 11) discuss, was consistently imbued with racial politics. Again critiquing the Spanish and their Old World mismanagement, American investor Charles Morris lambasted the ancient methods of sugar refining, which resulted in a "black, pasty mess," that were still in practice because of the "conservatism of the natives" and an apathy instilled by the Spanish so that Filipino landowners "do not care to change from the system first taught them."[17] A golden opportunity awaited the adventurous and modern American investor who would simply move to the Philippines. Rice farming too presented an ideal investment opportunity for applying American agricultural science, boosters thought. A 1903 pamphlet, *Modern Rice Culture*, forecasted immense profit for American investors. "There is no reason why scientific methods of culture and modern agricultural implements should not make the Philippines one of the leading rice countries of the world," it reported.[18] Such a transformation of rice farming could enrich Americans, but it could also generate goodwill in the islands by feeding Filipinos.

The American Chamber of Commerce provided additional encouragement for investing in Filipino food items. In one example, the Manila-based organization promoted aquaculture, a seemingly natural fit for a nation of islands, as an ideal opportunity to both feed Filipinos and enrich American investors. "There is ample opportunity for a great business in salt dried, smoked, and pickled fish," the chamber wrote, "for the first person who is willing to put up a clean, sanitary, wholesome product of uniform excellence." It too criticized the historical mismanagement of Spanish colonial rule in aquaculture, lamenting that Filipinos had become dependent on imported sardines and salmon instead of consuming fresh fish from around the country. The combination of American investment and Philippine natural resources would bring canned Indigenous fish to the masses: "We have the fish, we have an abundance of labor, we have unlimited quantities of vegetable oil, we have markets galore, then why don't we fish ourselves? . . . Manila should be supplying a large share of the canned, dried, and preserved fish these people consume, instead of allowing her fish to waste and importing to supply the deficiency."[19] Even in this case, when the purpose was not to promote export commodity markets, the actions were meant to cater to American investors first, with Filipino eaters as presumed beneficiaries. Through food, Americans could raise their new colony while turning a profit. It also promoted the Philippine cattle industry, arguing native breeds could reclaim the market from imported canned meat.[20] It pushed Philippine coffee and tea, two items that had languished under Spanish rule despite their long historical traditions. Even the smallest food items became potential moneymakers. The American Chamber of Commerce promoted palupo, a Philippine tea, by citing US Department of Agriculture scientist Dr. George T. Mitchell's optimistic forecast for this fringe crop's international consumption. Mitchell argued that "this Philippine tea might have a commercial future as a non-stimulating tea, on par with Postum and similar non-stimulating beverages." He envisioned factories of machine rollers and tea-drying machines producing palupo for Americans abroad as well as Filipinos.[21] Even the smallest agricultural commodities offered opportunities to make money and garner favor among Filipinos. Finally, it encouraged investors to collaborate with agricultural scientists already working in newly established research labs and agricultural science departments at the recently established University of the Philippines to develop new models for Philippine farming.[22]

Most importantly, the American Chamber of Commerce joined the chorus championing the modernization of the sugar industry. Even after thirty years of American rule, it complained that Old World thinking still afflicted Filipino landowners who were content with inefficient methods and low yields. The chamber criticized the typical Filipino sugar plantation owner as a "gentleman" who maintained an outdated worldview, remained unconcerned with agricultural productivity, and was more interested in his role as the patron of a small town:

He is hospitable to a guest, gay with women, paternal to subordinates; as a class he gambles inveterately, generously, and fairly; he neither abjures religion nor gives it much attention, but his children are born and reared in it, his dead buried in it; and he makes a thoroughly bourgeois marriage in which the ends of property and propriety are docilely and obsequiously subserved.[23]

Philippine sugar needed American assistance to free it from old habits and middling outputs. What is more, a sugar baron failed to think beyond his small town and would instead "fritter a fortune away on a single campaign for a provincial governorship, or the privilege of stepping from the gubernatorial chair to the Philippine legislature."[24] In contrast, placing Americans in charge would result in higher levels of prestige, opportunities for stockholding, and additional investments in sugar—central processing, steamship transport, and import-export companies.[25] With Yankee standards and guidance, the future of the Philippines would hereby rely on food and the American consumer market.

These calls to transform how Americans thought about Philippine food and agriculture ultimately made the Philippines into a place in which to exercise American benevolence, and travel guides directed visitors to specific restaurants to witness the fruits of their reforms. Steering tourists to areas where American investment flourished presented a Philippine narrative of improvement and showed food as a marker of progress. *Kemlein & Johnson's*, a 1908 publication for travelers in Manila, noted that the city's best hotels and restaurants cooked with ingredients such as potatoes, onions, cabbages, celery, apples, and cauliflower imported from Australia, Japan, and China.[26] *The Navy Guide to Cavite and Manila* in 1908 included a glowing account of Mrs. Smith's, an American restaurant in downtown Manila, that offered porterhouse and tenderloin steaks served by Chinese waiters in private rooms adorned with electric fans.[27] *Kemlein & Johnson's* also celebrated the ability to eat Chinese chop suey, Japanese sukiyaki, plain American dishes, French specialties, Spanish classics, banquets, and wine, but it failed to mention a single Filipino restaurant.[28] John D. Ford, an American engineer who traveled extensively throughout the Pacific, wrote in his 1898 travelogue *An American Cruiser in the East* that the French restaurant in Manila catered to "Spaniards and some natives of the better class . . . in an atmosphere already rich with garlic." He then provided an impressive list of Western-style foods such as "fine soup, fish, boiled potatoes, mystery, shrimp, salad, Spanish meat-balls, more mystery, capon and fried potatoes, claret ad libitum, assorted fruits, small cakes, ice cream, black coffee, and good cigars."[29]

Contrastingly, high-end Filipino cuisine, with its Hispanicized dishes and banquet service, failed to appear in these accounts. *You're in Manila Now*, a guide published by the US Navy, also included an impressive array of restaurants without a single Filipino establishment: Monsieur Savary's French Restaurant in Pasay, American southern cuisine at Tom's Dixie Kitchen in Santa Mesa Heights, Chinese food at Town Tavern in Plaza Santa Cruz, Italian food at Italian Restaurant in Dasmarinas, Spanish cuisine at Casa Curro at Lepanto Street, and Russian food at Continental in the Mabini-Ermita District.[30] The ability to eat international cuisine told travelers that the Philippines was indeed improving.

By the 1930s, these developments in food emerged as a key feature in the promotion of visiting the islands. The Philippine Bureau of Tourism created advertising campaigns that encouraged Americans to see how US investment had changed the country's food culture. They featured food alongside lodging, transportation, personal necessities, entertainment, and luxuries as the country's main highlights. They cited statistics showing the importance of food to travelers to the Philippines, as 41 percent of the average traveler's spending went to hotels and food.[31] The bureau encouraged travel agents in the United States to use food to promote the country's "unsurpassed" hotels and restaurants, arguing that these food establishments were larger draws than sports, leisure, low steamer rates, and tropical climate.[32] A well-fed traveler could return to the United States confident that the Philippines was headed in the right direction.

While simultaneously using food to justify American empire, Americans also used food to reinforce a racist hierarchy that denigrated Filipinos. The war and subsequent colonial occupation so many history textbooks gloss over were an era of redefining food and agriculture that created a modern form of colonized food. Food emerged as a preferred medium for reformers to uplift supposedly inferior peoples and transform Philippine society by changing how Filipinos ate. Introducing domestic science and changing how Filipinos prepared food meant that Americans could assert good intentions in healthier, cleaner, and safer eating. They also embraced modernizing Philippine agriculture as the primary engine for the country's economic revitalization.

This story of American empire and food took place more than a hundred years ago, but its ramifications for Filipino food's identity, particularly in the United States, resonates today. The relative absence of Filipino food in American popular culture is a direct result of efforts to denigrate Filipino cuisine in the age of American empire. Multiple foreign influences have shaped American cuisine, yet the small presence of Filipino food suggests that the United States has largely failed to confront its guilt-ridden, historically imperial relationship with the Philippines. Other former Western empires have wholeheartedly practiced culinary appropriation of the

Global South. Curry shops are staples of British town centers. The Dutch relish rijstaffel, or rice banquet tables, from Indonesia. The French have adopted Vietnamese bánh mì sandwiches. In contrast, the United States fails to acknowledge the food of its former colony in Southeast Asia. The act of consuming Filipino food raises questions about an imperial history that many Americans either do not know or would rather avoid.

Furthermore, the story shows how many Americans in the Philippines between 1898 and 1946 simply chose not to engage with Filipino food and culture. For most Americans in the islands the future looked American, with a distinct break from a Hispanicized or Indigenous past. Healthy Filipinos would eat American food, Filipino farms would produce for American consumers, and food would capture what was explicitly a one-way imperial relationship.

As a Filipino American and a former worker in the food industry, I wish I could tell my younger self that this is the reason there were (and are) so few Filipino restaurants in the United States to this day. Examining how food was a tool for historically exercising authority in the Philippines uncovers an uncomfortable but important lesson about the continuing legacy of American empire. American historical hostility to Filipino food shows racism's power in transforming daily lives and belittling Indigenous cultures. The accompanying push to transform Philippine agriculture reveals a fever for speculation that ignored the Philippine economic future and the nutrition of its people. For those Filipinos who emigrated to the United States in large numbers after the passage of the Immigration Act of 1965, these lessons of empire persisted so strongly that they remained reluctant to share their cuisine with Americans.

Today, Filipino food is belatedly making inroads as one of the next big ethnic cuisines. Adobo recipes appear in Instant Pot cookbooks; sisig (pork cheeks with chili and citrus juice) recipes appear as favorites at trendy urban food trucks. Philippine Brand dried mangoes are available at Costco. Having ignored Filipino cuisine for so long, many Americans are now turning to it with fascination, realizing that what they once considered uncivilized is now suddenly cosmopolitan. But we must not forget that, as recently as sixty years ago, even American writers, scientists, investors, and bureaucrats living in the Philippines did not want to consider Filipino food at all; they simply wanted an American colony ready to submit to American hands. Indeed, Filipino food itself was an afterthought in the larger mission of benevolent uplift. But the attempt to transform it and make it more American was, in short, a cornerstone of American empire and the beginnings of the American conception of Filipino food today.

NOTES

1. Caroline S. Shunk, *An Army Woman in the Philippines: Extracts from Letters of an Army Officer's Wife, Describing Her Personal Experiences in the Philippine Islands* (Kansas City, MO: Franklin Hudson Publishing Co., 1914), 58.

2. Jacob Isselhard, *The Filipino in Every-Day Life: An Interesting and Instructive Narrative of the Personal Observations of an American Soldier during the Late Philippine Insurrection* (Chicago: The Author, 1904), 27.

3. Joseph Earle Stevens, *Yesterdays in the Philippines* (New York: Scribner's, 1898), 25.

4. Stevens, *Yesterdays in the Philippines*, 150–151.

5. José de Olivares, *Our Islands and Their People as Seen with Camera and Pencil* (New York: N. D. Thompson, 1899), 553.

6. E. W. Stephens, *Around the World: A Narrative in Letter Form of a Trip around the World from October, 1907 to July, 1908* (Columbia, MO: E. W. Stephens, 1909), 176.

7. B. J. Kendrick, "American Who Made Health Contagious," *Harper's*, April 1916.

8. John Clifford Brown, *Diary of a Soldier in the Philippines* (Portland, ME: Lakeside Press, 1901), 178.

9. John M. Lewis, "Letter to Sophie Borel, 20 April, 1900," Correspondence, Bancroft Library, Berkeley, CA, 1898–1915.

10. Elsie McCloskey Gaches, *Good Cooking and Health in the Tropics* (Manila: American Guardian Association, 1922), 300.

11. George Kindley, "The Lumbayo Settlement Farm School," *Philippine Craftsman* 2, no. 8 (February 1914): 564.

12. Genoveva Llamas, "How the Teaching of Domestic Science Is Influencing the Home," *Philippine Craftsman* 4, no. 8 (February 1916): 525; Elvessa A. Stewart, *Foods in Relation to Health*, Philippine Public Schools, November 1929.

13. Arthur David Hall, *The Philippines* (New York: Street and Smith, 1898), 43.

14. Hamilton Wright, "Unknown Philippines," *World To-Day*, December 1907.

15. Theodore W. Noyes, Conditions in the Philippines: June 4, 1900, Ordered to Be Printed, Mr. Morgan Presented the Following Editorial Correspondence of the Evening Star, Washington, D.C., 56th Congress, 1st Sess., Senate, doc. 432 (Washington, DC: Government Printing Office, 1900), 61–62.

16. O. W. Barrett, *The Philippine Coconut Industry*, bulletin no. 25 (Manila: Bureau of Printing, 1913), 9, 64.

17. Charles Morris, *Our Island Empire: A Handbook of Cuba, Porto Rico, Hawaii, and the Philippine Islands* (Philadelphia: J. B. Lippincott, 1906), 447–448.

18. W. J. Bodreau, *Modern Rice Culture*, bulletin no. 3 (Manila: Bureau of Printing, 1903), 22.

19. Albert W. C. T. Herre, "Aquatic Resources of the Philippines," *American Chamber of Commerce Journal*, September 1921, 11–12.

20. "Trends in the Philippine Cattle Industry," *American Chamber of Commerce Journal*, September 1932, 3.

21. Roberto Francisco, "Franciscan Friar Brings Coffee to the Philippines," *American Chamber of Commerce Journal*, May 1928, 9.

22. Francisco, "Franciscan Friar Brings Coffee to the Philippines," 7.

23. "Cane Sugar in the Philippines," *American Chamber of Commerce Journal*, August 1931, 6.

24. "Cane Sugar in the Philippines," 6.

25. "Cane Sugar in the Philippines," 6.

26. H. Kemlein, *Kemlein & Johnson's Guide and Map of Manila and Vicinity: A Hand Book Devoted to the Interests of the Traveling Public* (Manila: Kemlein & Johnson, 1908), 56.

27. *Navy Guide to Cavite and Manila: . . . A Practical Guide and Beautiful Souvenir* (Manila: n.p., 1908), 50–51.

28. Kemlein, *Kemlein & Johnson's Guide and Map of Manila and Vicinity*, 52.

29. John D. Ford, *An American Cruiser in the East: Travels and Studies in the Far East; the Aleutian Islands, Behring's Sea, Eastern Siberia, Japan, Korea, China, Formosa, Hong Kong, and the Philippine Islands* (New York: A. S. Barnes, 1898), 431.

30. United Service Organizations Incorporation. *You're in Manila Now: Information for Service Wives* (Manila: United States Army, n.d.), 49.

31. Thomas Cook Ltd., *Information for Travellers Landing at Manila* (Manila: T. Cook & Sons, 1913), 67.

32. James King Steele, "Values of the Tourist Industry to the Community," 1934, James King Steele Papers, 1909–1936, Huntington Library, San Marino, CA.

4

Modern Food as Globalized Food

DOES YOUR BEER HAVE STYLE?
THE NINETEENTH-CENTURY INVENTION
OF EUROPEAN BEER STYLES

Jeffrey M. Pilcher

Long derided as an insipid, proletarian beverage, beer is now considered stylish, and not just among those with a penchant for groomed beards and vintage dresses. Aficionados have lately grown dissatisfied with the bland, industrial, pilsner-style beers distinguishable only by brand names such as Budweiser, Corona, and Heineken. Instead, beer seekers like me prefer craft beers with recognizable origins and distinctive flavor profiles: the dark chocolate and coffee of London porter, the bitterly hoppy citrus of India pale ale, the amber color and floral notes of Vienna lager, and— the Holy Grail of craft beer enthusiasts—the barnyard funk of Belgian sour ales.

Why pursue such acquired tastes when pilsners are cheap and refreshing? Critics have derided the fad as a white male status symbol, popular among Brooklyn hipsters whose careers as corporate executives and wine snobs were derailed by the financial crisis of the 2000s. Craft brewers, by contrast, trace their movement back to the 1960s countercultural revolt against globalized, industrial food production. As corporations seduced consumers with the convenience of a global diet with fast food and year-round produce, dissidents sought to reclaim the pleasures of home cooking with local, seasonal fruits and vegetables. Craft beer likewise offered an opportunity to connect with seemingly more traditional, authentic, and tasty food and drink.

Craft beer's foremost advocate, the British journalist Michael Jackson, became known to followers as the "Beer Hunter." With a mop of curly hair, rumpled tweed jacket, and a tulip glass of Belgian ale, he looks drunk in many photos, although his unfocused affect was a symptom of Parkinson's disease, from which he died in 2007. From the 1970s he had guided readers to surviving local breweries while encouraging the nascent craft beer movement to revive old favorites. On his website he declared: "The last major style of beer to be introduced was Pilsner in 1842, but in all styles there are new efforts to rival the classics."[1]

Defining traditional beer styles was a vital part of Jackson's evangelizing mission. Since few people had experienced many local brewing traditions, Jackson sought to guide efforts at historic preservation. He constructed a Linnean taxonomy of beer styles, which divided beers into categories based on their ingredients, brewing methods, and distinctive flavor profiles. As Garrett Oliver observed in the *Oxford Companion to Beer*, "Although we tend to imagine that the modern concept of beer style is itself ancient, it is not. In fact, it is not even old, having been essentially invented out of whole cloth by the late beer writer Michael Jackson in his seminal 1977 book *The World Guide to Beer*."[2]

Oliver exemplifies the cosmopolitanism present in the best of contemporary craft brewing. In his day job as brewmaster at the acclaimed Brooklyn Brewery he produces prizewinning beers in a wide range of styles, but he also travels the world in his trademark straw hat and scarf, holding tasting seminars and collaborating with like-minded brewers around the world. On one trip to Brazil he brewed beer with the juice of freshly cut sugarcane. His experimental beers, such as Improved Old Fashioned, a barrel-aged rye ale with bitters, and Cuvée Noir, a stout with Mauritius sugar and orange peel, have defined a personal style while remaining faithful to the basic categories surveyed by Jackson.[3]

The craft beer movement thus shares a now-common impulse to revive old traditions, to categorize them, and to create innovative twists on them. But in our nostalgia we sometimes forget that brewers of the past could likewise be innovators and not just follow traditions that had supposedly persisted from time immemorial. We also risk erasing the fact that industrial beer, like most other modern food, arose in tandem with efforts to invent white European identities. Many of the classic beer styles that inhabit Jackson's taxonomy were, like pilsner, artifacts of nineteenth-century capitalism. Brewers in cities such as London, Munich, Vienna, and Pilsen, Bohemia, cultivated distinctive styles in order to survive in an increasingly competitive market but nevertheless industrial consolidation overwhelmed countless local firms. Pilsner triumphed over these rival styles precisely because of its blandness, which consumers equated with nineteenth-century ideals of purity and modernity. Likewise, many now fear for the sustainability of today's golden age of craft beer, which has been challenged externally by transnational giants and internally by a culture of masculine exclusivity and the excessive pursuit of novelty. Understanding the capitalist forces that shaped beer styles in the nineteenth century may help craft brewers survive our own cutthroat market while also encouraging beer drinkers to make more modest identity claims based on their consumption choices.

For thousands of years, brewers around the world fermented grains and other starches, defying the consistency we expect from modern beer styles. These brewers, overwhelmingly women, varied their recipes according to the staples available from fields or markets, the personal selection of myriad flavorings and preservatives, the source of water, and the unseen vagaries of yeast, both familiar strains that they nurtured in pots from one batch to the next and wild interlopers that could contaminate and ruin their work. This domestic economy of brewing became commercialized in early modern Europe. The introduction of hops made it possible to brew on a larger scale and to keep beer for longer than a few days without it turning sour, encouraging the growth of international markets. To maximize profits, male-dominated guilds sought to marginalize the subsistence production of women. Although the industrialization of the food supply created space for prominent female participation, as described in Adam Shprintzen's essay on Ella Eaton Kellogg (chapter 13), in this and other cases it also helped eliminate female-dominated modes of production. Ultimately, the rise of industrial brewing in eighteenth-century Britain and consumer expectations for a reliable product drove the standardization of beer into styles recognizable today.

According to legend, the first industrial beer style, London porter, was invented in Shoreditch, London, in 1722 by a brewer named Harwood.

Tavern-goers of the period often requested a pint poured from three distinct casks. Harwood's inspiration was to combine all three varieties into a single cask, known as "entire butt," thus saving the barman the trouble of juggling multiple taps. This full-bodied brown ale, aged for a "stale" but not sour taste, became a favorite among London market porters, hence the name. It could also be produced easily in enormous vats, and over the course of the eighteenth century a handful of brewers achieved economies of scale in supplying porter to London's burgeoning industrial workforce.[4]

Nevertheless, the reign of porter was short lived, as new technologies made it possible to brew pale ale on an industrial scale, avoiding the inefficiencies of dark malts, the roasting of which burned away fermentable matter along with profits. In the 1810s, with parliamentary approval, London brewers began brewing ale and transforming it into porter by adding caramelized sugar. The public was coming to prefer the bright appearance and mellow flavors of pale ale, and so imperiled London brewers undertook what the historian James Sumner has called the "retrospective invention" of porter, cultivating its creation myth along with the now-characteristic black color, made possible by the use of caramel, as a form of product differentiation. Their efforts failed, and the porter market passed to a Dublin firm called Guinness.[5]

Romantic tales of India pale ale, supposedly created to refresh colonial officials in the tropics, likewise served primarily as advertising for the home market. As the classic origin story goes, a London brewer named George Hodgson brewed the first pale ale for the East India Company in the late eighteenth century, using higher levels of alcohol and an extra strong dosing of hops to provide stability for the oceanic voyage. By the 1820s, as the British extended their control over South Asia, brewers such as Allsopp and Bass, from the town of Burton-on-Trent, cut into the colonial market. With the growth of Britain's railroad network in the 1830s, Allsopp and Bass also began to compete in London, to the dismay of the capital's porter brewers. But this origin story also lacks historical perspective. Brewing stronger and more heavily hopped ales for stability was standard practice among Burton firms exporting to Baltic countries throughout the eighteenth century. Moreover, the India trade amounted to scarcely a drop in the proverbial barrel compared to Britain's domestic market. Images of colonial troops in pith helmets quaffing India pale ale helped foster the self-image of white English colonialists, mostly satisfying the imperial nostalgia of its primary consumers, London clerks and shopkeepers.[6]

Modern beer is global beer, and the nostalgic mythmaking around products of a capitalist marketplace extended past eighteenth-century England. The development of Central European beer styles was influenced by the rise (and fall) in popularity of lager beer from the Bavarian capital, Munich. Writing in 1867, an Austrian industrial inspector named J. John described this *altbayerische Art* (old Bavarian style) as a "brown beer with seemingly stronger concentration brewed from dark malt and with an unusually long cooking of the hopped wort."[7] Although often attributed to the Reinheitsgebot (purity decree) of 1516, crucial elements of the *altbayerische Art* were scarcely thirty years old at the time. The man often credited with wrenching the Bavarian brewing guild into the modern, industrial era, Gabriel Sedlmayr Jr., left Munich in the late 1820s on a journeyman's tour of Europe. In England, along with a Viennese colleague named Anton Dreher, Sedlmayr undertook a campaign of industrial espionage. Using hollow walking sticks equipped with valves, they stole and analyzed samples from leading breweries in London, Burton, and Edinburgh. Sedlmayr wrote: "It always surprises me that we can get away with these thefts without being beaten up."[8] Returning home to Munich, he incorporated the new technology at the family firm, Spaten, which became the city's leading brewery and a model for all of Europe.

Bavarian lager beer resulted from a unique variety of bottom-fermenting yeast. Unlike the ubiquitous, top-fermenting *Saccharomyces cerevisiae*, which has been used for millennia to ferment ale, bread, and wine, the hybrid *Saccharomyces pastorianus* was tamed in Central Europe during the early modern era. When both strains of yeast were present in the brewery, the resulting beer depended on the temperature of fermentation. *S. cerevisiae* works best between 18° and 22° Celsius, while *S. pastorianus* ferments more slowly but remains active at temperatures as low as 5° to 10° Celsius. Early modern brewers thus faced not an ale/lager dichotomy, but rather a continuum defined by temperature. Most often, *S. cerevisiae* reproduced quickly enough to take control of the reaction, but its sluggish start when cold, such as during a Bavarian winter, could allow *S. pastorianus* to flourish. A change in the weather could even reverse the balance within an individual batch of wort, turning ale into lager, or vice versa.[9]

Bavarian brewers took the lead in developing bottom fermentation as a result of another sixteenth-century beer law, which prohibited brewing with barley between St. George's Day (April 23) and St. Michael's Day (September 29). Ostensibly intended to ensure quality, since summer brewing risked runaway fermentation and wild yeast contamination, the edict also increased demand for *weissbier* (wheat beer), a monopoly on which was held by the Duke of Bavaria. To sell beer during the thirsty summer season, guild members brewed stronger beers over the winter and placed them in storage chambers (*lager*, in German), generally ice-filled caves where *S. pastorianus* worked its slow magic. The refreshing *sommerbier* or *lagerbier*, served chilled from the lager to sweaty customers, won fame for Bavarian brewers already in the eighteenth century.

Although bottom fermentation was well developed by the early nineteenth century, Munich lager was not, as a rule, *dunkel*, the dark-brown color that John and others had come to expect by the 1860s. Indeed, consistency of color as a marker of style was an important, although overlooked, British innovation that Sedlmayr helped introduce to continental Europe. In 1829, as the young journeyman was setting out on his tour, the *Wöchentlicher Anzeiger für Biertrinker* (Weekly Beer Drinker's Gazette) reported on a sample of Munich beers including twenty-eight *weingelb* (wine yellow), twenty-two *hellbraun* (light brown), and only one *dunkelbraun* (dark brown).[10] In 1835 Sedlmayr and Dreher separately conducted their first experiments in brewing Bavarian beer with English pale ale malt.[11] Sedlmayr dubbed this amber-colored beverage *Märzenbier* (March beer), a nostalgic reference to strong lager that was stored through the summer for consumption in early fall, especially at Oktoberfest, itself a recently invented tradition celebrating the marriage of Bavarian Prince Ludwig and Princess Therese in 1810. For his signature "Munich" beer, Sedlmayr set a more ambitious and technically challenging goal, to brew a full-bodied dark beer without the British trick of adding caramel sugar, which was forbidden in Bavaria. By replacing cantankerous, smoky malt ovens with the indirect and carefully regulated heat of British flue-kilns, it became possible to toast the malt for about a day "at a gentle, clean heat, without being browned in the slightest degree," and at the very end to "suddenly raise the temperature, which brings out the semi-caramelized substance, believed to explain the peculiar richness and aroma of the beer."[12]

Bavarian dark lager beers became wildly popular across Germany in the 1840s and 1850s. Years later the brewmaster Ernst Rüffer sardonically recalled taverns, "especially in little country towns (*Landstädtchen*), under the name of 'Bavarian Beerhall,' cheerfully visited by a discriminating clientele, who preferred the 'genuine Bavarian' and so-beloved *Doppelbier*; many enjoyed it simply because it belonged to 'high society.'"[13] But fashion alone was not the only appeal; consumers generally believed lager beer, like many novel industrial foods of the nineteenth century, to have a fresher taste and more hygienic character than often sour or contaminated top-fermented ales. The type of beer one drank could indicate hierarchy, as could the white bread that David Fouser describes in chapter 2. Especially in the "hunger years" of the 1840s, as the potato blight raged out of Ireland and across Europe, a dark, full-bodied, double-strength beer offered a satisfying symbol of wealth and status.

A globalizing world helped spread lager, even as it posed challenges to its producers. Munich's connection to the German railroad network in 1840 opened new markets, but rival brewers quickly adopted bottom fermentation and began to compete for the "Bavarian" beer market. A certain Herr Goschenhofer had reportedly opened Berlin's first Bavarian-style brewery in 1829, and a few years later the Leipzig cloth merchant Maximilian Speck von Sternburg traveled to Munich to obtain the latest lager brewing technology, including plans for the newly constructed Augustiner Brewery. In towns throughout Germany, as Rüffer observed, the fashion for lager "forced brewery owners to abandon top fermentation and make dark, bottom-fermented beers with the same [Bavarian] taste and characteristics."[14]

The dilemma for exporters like Sedlmayr was thus the replicability of the recipe for lager. The "genuine Bavarian" article retained its appeal, and by 1870 the southern kingdom produced more than 10 million hectoliters of beer, one-third of the output for the entire German Empire—and not just because the people of Munich drank five times the national average. Meanwhile, Sedlmayr channeled his youthful masculinity, risking violence by stealing beer samples, into building Munich's leading brewery.[15] But notwithstanding his influence, he was only one among many innovators at the time. Moreover, as secretive guild practices gave way to an open exchange of scientific knowledge, the "industrial espionage" of his British mission—or, indeed, the contemporary travels of Speck von Sternburg—acquired the more neutral term "technology transfer."[16] Even Sedlmayr's triumph of associating the full-bodied, bottom-fermented Bavarian lager with a distinctive dark color left a mixed legacy. As London's porter brewers had unhappily discovered, fashion was fickle and upstarts eagerly challenged market leaders. Moreover, innovations in one place led to imitation and competition elsewhere.

As in Germany, the rise of industrialization, urbanization, and railroad transport created the conditions for growing consumer markets for brewers in the Austro-Hungarian Empire. Manufacturers here, and particularly in the Bohemian lands of the present-day Czech Republic, created some of the most popular beer styles. Inspired by lager, towns such as Pilsen sought out Bavarian experts to improve the quality of local production. In other cases, notably Anton Dreher of Vienna, brewers ventured out to learn new technologies. Their breweries became laboratories of experimentation in which modern stylistic forms took shape in competitive international markets.

Few visitors in the early nineteenth century might have guessed that Pilsen, a market town on the road from Bavaria to Prague that had lately grown rich in the grubby business of anthracite mining, would become world renowned for its clear, sparkling beers. The local private and monastic breweries were losing customers to imported Bavarian lagers, and the town fathers decided to invest some of the profits from mining into the construction of a municipal brewery (*Bürgerliches Brauhaus*). As Michael Jackson suggested, the original batch of Pilsner became a modern legend: on October 5, 1842, the Bavarian brewmaster Josef Groll mashed in golden malt from Moravian barley with the town's soft spring water, added aromatic Saaz hops from the Žatec Basin north of Pilsen, and fermented the wort with Bavarian lager yeast. Tasting the new beer on November 11, the burghers of Pilsen heartily applauded Groll's work.[17]

Apart from this basic recipe, we know little about how pilsner acquired its legendary status. It was valued, at first, as an exotic "Bavarian" beer, and the malting process relied on British advances in indirect heating, like Sedlmayr's Munich malt, but without the final burst of caramelization. The fermentation vat held only thirty-six hectoliters (about twenty barrels), and the Bürgerliches Brauhaus sold primarily to local markets. Even after the construction of an expanded brewhouse in 1852, production grew slowly until the opening of the Böhmische Westbahn (Bohemian Western Railroad) in 1862 brought steady demand from Prague, Vienna, and abroad. In 1868, the *Bayerische Bierbrauer* (Bavarian Brewer) noted the appeal of the Austro-Hungarian Empire's beers, attributing their popularity to the qualities of "lightness, carbonation, mild taste and clear golden color." The author listed Pilsner as the first among several Bohemian export beers, along with Egerer and the Prague-based Kreuzherrenbier, Königsfaaler, Nussler, and Bubner, although curiously not Budweiser, which was first brewed with Bavarian bottom fermentation in 1852.[18]

In any event, midcentury observers were far more impressed by Klein Schwechat, the Vienna brewery of Anton Dreher, partner to Gabriel Sedlmayr. Both had experimented with combining English pale ale malts and Bavarian lager yeast, but whereas Sedlmayr reserved the recipe as a seasonal specialty, Märzenbier, Dreher used the amber brew as the model for his signature Vienna lager. The Austrian capital swooned for its namesake beer, and workers rushed to keep up with demand from the rapidly growing population. From an initial output of 16,000 hectoliters in 1836–1837, production rose above 50,000 hectoliters by 1848, then tripled again in the following decade, making "Little Schwechat" the biggest brewery in continental Europe. Nor was that enough for the hyperactive Dreher, who opened two new breweries to capture regional markets: Michelob in the Saaz district of Bohemia, in 1861, and Steinbruch in the Hungarian capital of Budapest, in 1862. Exhausted from the pace, he died on December 26, 1863.[19]

In the case of Vienna lager, and probably pilsner as well, there was no primordial recipe, perfected on the first try, but rather an ongoing interaction between producers and consumers, ingredients and technology. The Frankfurt brewmaster Friedrich Henrich recalled working as a journeyman for Dreher in 1858: "At the time he was famous as a maker of Märzenbiere"—as Vienna lager was often called in Germany—"and it was his wish that this Märzenbiere should be more full-bodied than it had been previously, and particularly that it should be different from highly fermented (sehr vergohrenen) lager beers."[20] It is unclear whether Henrich referred here simply to spoiled beer or to highly attenuated, drier varieties, common to Bohemia, in which the sugars had been converted more thoroughly to alcohol. Regardless, the remark indicates Dreher's persistent concern to refine his brewing style—and with an eye to the competition. Likewise, in the 1890s, Bohemian brewers concluded that they had been kilning their malt at too low a temperature and intensified the roast to add more flavor.[21]

By the 1860s, Central European brewers had created a spectrum of lager beers ranging from dark, full-bodied Munich to amber, malty Vienna and golden, dry pilsner—and that was no accident. Even before the railroads had fully integrated European beer markets, brewers kept a close watch on their rivals to make their own products stand out for consumers. Such attention to markets became increasingly vital as international competition intensified.

The final third of the nineteenth century was a golden age for European beer drinkers and a Darwinian struggle for survival among brewers. Technological advances in railroads, refrigeration, and bottling encouraged the shipping of premium beers across the continent, giving consumers more choices. Meanwhile, in many formerly wine-consuming regions such as the Mediterranean and the Balkans, peasants who had migrated to cities in search of factory work acquired a taste for industrial beer. But whether in traditional beer-drinking countries like Germany and Britain or in newcomers like France and Italy, consumers increasingly chose light, clear, blond beers like pilsner, making this literally a "golden age." Nevertheless, as in the midcentury "dark age" of Munich dunkel, local brewers challenged the "genuine" article from Pilsen.

Production and trade statistics provide one indication of the growth and integration of Europe's beer market in the late nineteenth century. Exports from the Habsburg Empire to Germany took off in 1865, when heavy tariffs were repealed, reaching 400,000 hectoliters by the 1890s. Bavarian brewers meanwhile shipped more than 1 million hectoliters a year, mostly to other German states, but also to foreign countries, especially France and Switzerland.[22] But notwithstanding this export boom of premium lagers, most beer was consumed locally because of high shipping costs for what was essentially water.

As producers fought for consumers, they built global brand names. The product of efforts to make beer brewing seem a white male occupation, modern beer makers advertised it as a modern, urban drink. Promotional literature often bore pristine alpine vistas to evoke the purity of ingredients, but beer labels and postcards also commonly showed smoke-billowing factories to demonstrate the modernity of production.[23] Even beer gardens, with servers in folkloric peasant dresses (dirndls), marketed agrarian nostalgia to urban dwellers. Other important sites for globalizing brands were world's fairs, those showcases of nineteenth-century industrial modernity. North American brewers fought for bragging rights at the 1893 World's Columbian Exposition, where, in a controversial reading of the judges' scoring cards, Pabst declared victory over Anheuser-Busch and awarded itself a "Blue Ribbon."[24] Decades earlier, an eighteen-year-old Anton Dreher Jr. had proved himself a worthy heir to the family firm through a charismatic performance at the 1867 Universal Exposition in Paris. From a kiosk at Châtelet, the tall young brewer handed out miniature glasses of beer drawn from kegs that had been carefully shipped from Vienna through the summer heat in novel, ice-chilled railroad cars. A German observer later recalled: "To drink a 'bock' (goat), as the Parisians called such beer glasses, with Dreher became a beloved and overwhelming fashion in Paris."[25]

But Vienna lager's conquest of Paris in 1867 was simply a prelude to the continental hegemony of pilsner. On the eve of the Universal Exposition, the Austrian industrial official John declared: "In the struggle between light and brown beer, it appears that the light is gaining more followers day by day."[26] Dr. Carl Lintner, who published (or perhaps reprinted) the article in the *Bayerische Bierbrauer*, added a footnote describing the appeal of full-bodied Bavarian beers in greater nourishment, but it was a bad sign that he felt the need to explain. By 1873 the Bavarian journal declared that pilsner was "preferred to the famous Viennese beers, even in Vienna."[27] A decade later the Böhmische Brauhaus (Bohemian Brewery) had become Berlin's leading producer, and demand for pilsner beer had spread across northern Germany.[28]

The simultaneous zenith in Bavarian beer exports and eclipse in the popularity of Munich lager raises an apparent contradiction, but an easily resolved one. John cautioned brewers "not to throw ourselves exclusively into a single type of beer, but rather we must study the needs and tastes of other countries and direct our production accordingly."[29] Setting profits over pride, Bavarian brewers followed his advice and adapted their output to foreign conditions. In particular, by brewing in the style of Vienna, they stole the Parisian market out from under Dreher, even after the Franco-Prussian War of 1870 made hostility to all things German a patriotic duty for the French.[30]

Ultimately, the preference for pilsner and similar beers—light, clear, mild, fizzy, and low in alcohol—reflected a modern zeitgeist. Low alcohol content made it easier for industrial workers to quench their thirst without risking accidents, while also deflecting the ire of the growing international temperance movement. Moreover, well-fed urban consumers no longer needed the calories of the full-bodied Munich lagers and London porters that had nourished them through earlier lean years. Instead, they came to prefer a light beverage that did not slow them down in sports, dancing, and other leisure activities. Finally, mild flavor and clear appearance exemplified the hygienic promise of nineteenth-century food processing; or to put it another way, the pungent smell and taste of wild yeast and strong hops that today seem authentic and natural were perceived at the time as distasteful and potentially hazardous defects resulting from antiquated brewing methods.[31] Standardized pale beer indicated the heights of modernity.

Behind the pleasures of the consumer society lay structural changes that threatened this golden age. Mergers and bankruptcies were consolidating the industry, as the Berlin-based *Wochenschrift für Brauerei* (Brewers' Weekly) demonstrated with its 1888 plea: "No new corporate breweries!"[32] Workers as well as owners felt the squeeze as mechanization replaced jobs, while industrial accidents spread fear in the brewery.[33] Consolidation and scientific brewing schools culminated the centuries-long trend of excluding women from management in what had remained largely family businesses through the mid-nineteenth century. Women continued to work in isolated corners of the industry such as bottling and, more recently, quality control, but only with the rise of the craft beer movement have women returned to the brewhouse.[34]

By the end of the nineteenth century an international system of beer styles had been firmly established based on patterns of color, alcohol, and yeast rather than on local origins. Of these new styles, pilsner was already the leader, and even in Bavaria, which imported only miniscule quantities of beer, consumers increasingly preferred a light, blond beer called "helles" to distinguish it from the Bohemian competitor. Perhaps the biggest loser in the international competition was Austria. Not only had pilsner crowded the imperial capital's brewers out of export markets, but even the Vienna-style malt was disappearing throughout Europe.[35]

"That mysterious legend woven around the *old Pilsner beer*," declared Karl Dubský in 1899, "was nothing more than the exemplary purity of taste that put it at the forefront of all other beers."[36] Whereas today's consumers often take the purity of mass-market beers for granted and seek out unique characteristics in artisanal brews, overcoming sour tastes and dangerous contamination represented one of the great triumphs of nineteenth-century industrial food production. And even while celebrating "our modern beer," the Bohemian brewmaster Dubský was enough of a modernist to recognize that modernity lay not in a particular style but rather in the choice between them, which allowed consumers to express their individual identity and social distinction. "Where would our brewing industry be today, if it had only *one* taste to satisfy; so every taste finds its agreeable beer and every beer its admirer—the Pilsner like the Munich, the Culmbacher like the Gose."[37]

The nineteenth-century pursuit of novelty and distinction, by both brewers and consumers, in turn reflected the spirit of rootless industrial capitalism, and it persists today in the craft beer movement. Just as Gabriel Sedlmayr and Anton Dreher crafted their distinctive Munich and Vienna lagers through a conscious strategy of market segmentation, contemporary brewers seek to attract consumers with exciting new tastes. The experience of early industrial brewers, particularly the wave of concentration at the end of the nineteenth century, therefore holds lessons for today's microbrewers, who worry that their own business model may become unsustainable. While touting ties to local communities and even to some locally sourced ingredients, most breweries purchase malt and hops from transnational firms. Marketing often depends on distributors beholden to AB Inbev and other industry giants, who have muscled their way into the craft market. Although owners may draw a living wage, workers often scrape by on poorly paid internships in the hopes of acquiring the skills to open a microbrewery of their own.[38] Consumers have been willing so far to pay a premium for the quality and community of craft beer, but their continued dedication will ultimately determine whether today's microbreweries, like their counterparts a century ago, are forced to "get big or get out."

Indeed, consumers helped to inflate a craft beer bubble with their incessant demands for novelty and exclusivity. The insistence on finding a new beer to sample on each visit requires brewpubs to maintain impossibly long lines of taps. Moreover, the craft movement has cultivated a snobbish masculinity with undrinkably bitter IPAs that snarl, in the words of the heavily muscled demon mascot of Stone Brewery's Arrogant Bastard Ale, "You're Not Worthy." Adding context to beer culture, then, means undoing one of its founding premises: the cultural associations with white men. There are successful Black and female brewers, such as Garrett Oliver and Teri Fahrendorf, founder of the Pink Boots Society, which is dedicated to supporting women in the brewing profession. This is despite the roadblocks inherent to craft beer culture, which is based on a Eurocentric nostalgia and the exclusive actions of earlier brewers. Knowing the history of craft beer is one step toward changing that culture for the better, and so helping to sustain a golden age of brewing.

NOTES

1. "Tasting Notes," The Beer Hunter, accessed February 7, 2018, http://www.beerhunter .com/tastebystyle.html.

2. Garrett Oliver, "Beer Style," in *The Oxford Companion to Beer*, ed. Garrett Oliver (New York: Oxford University Press, 2012), 115.

3. "Garrett Oliver," Brooklyn Brewery, accessed July 1, 2018, http://brooklynbrewery .com/about/the-brewmaster.

4. Michael Jackson, *The World Guide to Beer* (Philadelphia: Running Press, 1977), 156.

5. James Sumner, "Status, Scale, and Secret Ingredients: The Retrospective Invention of London Porter," *History and Technology* 24, no. 3 (September 2008): 289–306.

6. Malcolm Purinton, "India Pale Ale" (unpublished manuscript); Alan Pryor, "Indian Pale Ale: An Icon of Empire," in *Global Histories, Imperial Commodities, Local Interactions*, ed. Jonathan Curry-Machado (London: Palgrave Macmillan, 2013), 38–57.

7. J. John, "Bier und Malzfabrikation in Oesterreich vom Standpunkte des Exportes," *Der Bayerische Bierbrauer* 2, no. 2 (February 1867): 22.

8. Quoted in Michael Jackson, "The Birth of Lager," The Beer Hunter, March 1, 1996, http://www.beerhunter.com/documents/19133-000255.html.

9. Holger Starke, *Von Brauerhandwerk zur Brauindustrie: Die Geschichte der Bierbrauerei in Dresden und Sachsen, 1800–1914* (Cologne, Germany: Böhlau Verlag, 2005), 47.

10. Christian Schäder, *Münchner Brauindustrie, 1871–1945: Die Wirtschaftsgeschichtliche Entwicklung eines Industriezweiges* (Marburg, Germany: Tectum Verlag, 1999), 97.

11. Wolfgang Behringer, *Die Spaten-Brauerei, 1397–1997: Die Geschichte eines Münchener Unternehmens von Mittelalter bis zue Gegenwart* (Munich: Piper, 1997), 166–167.

12. *One Hundred Years of Brewing* (1903, reprint; New York: Arno Press, 1974), 61–62.

13. Ernst Rüffer, "Ein Wort über das Enstehen und die Fernhaltung des Pechgeschmackes im Biere," *Allgemeine Brauer- und Hopfen Zeitung* 44 no. 188 (August 12, 1904): 2221.

14. Rüffer, "Ein Wort über das Enstehen und die Fernhaltung des Pechgeschmackes im Biere." See also Schäder, *Münchner Brauindustrie, 1871–1945*, 164–165; *One Hundred Years of Brewing*, 680; Starke, *Von Brauerhandwerk zur Brauindustrie*, 134–135.

15. *Statistik des Deutschen Reiches für das Jahre 1882* (Berlin: Verlag von Puttkammer & Mühlbrecht, 1882), 220; "Bierbrauerei in Oberbayern," *Allgemeine Brauer- und Hopfen Zeitung* 26, no. 61 (May 23, 1886): 706.

16. Behringer, *Die Spaten-Brauerei, 1397–1997*, 135, 167.

17. Jackson, "The Birth of Lager."

18. "Etwas Cultur-historisches und Statistisches vom Bier," *Der Bayerische Bierbrauer* 2, no. 8 (August 1867): 118; "Allgemeine Landes-Jubiläums-Ausstellung in Prag," *Der Böhmische Bierbrauer* 18, no. 13 (July 1, 1891): 293–300.

19. "Entwicklung und Stand der Dreher'schen Brauereien," *Allgemeine Hopfen Zeitung* 12 (February 13 and 18, 1868), 46, 50.

20. "8. ordentliche Generalversammlung," *Wochenschrift für Brauerei* 7, no. 26 (June 27, 1890): 637.

21. "Das Darren des Malzes," *Der Böhmische Bierbrauer* 26, no. 1 (January 1, 1899): 1–5.

22. "Bierbrauerei in Oberbayern," 706; W. May, "Statistisches," *Zeitschrift für das Gesammte Brauwesen* 17, no. 8 (1894): 64, 209.

23. See the advertisements in Behringer, *Die Spaten-Brauerei, 1397–1997*, 368.

24. Maureen Ogle, *Ambitious Brew: The Story of American Beer* (Orlando, FL: Harcourt, 2006).

25. "Deutsche Exportbiere," *Allgemeine Brauer- und Hopfen Zeitung* 22, no. 5 (January 15, 1882): 36.

26. John, "Bier und Malzfabrikation in Oesterreich," 23, 24.

27. "Die Fortschritte im Brauwesen auf den hochfürstl. Schwarzenberg'schen Gütern in Böhmen," *Der Bayerische Bierbrauer* 8, no 7 (July 1873): 104.

28. "Bierbrauerei in Berlin," *Allgemeine Brauer- und Hopfen Zeitung* 22, no. 90 (November 9, 1882): 821; Wilhelm Windisch, "Pilsener Bier—norddeutsches helles Bier," *Wochenschrift für Brauerei* 14, no. 30 (July 23, 1897): 377.

29. John, "Bier und Malzfabrikation in Oesterreich," 25.

30. Schäder, *Münchner Brauindustrie, 1871–1945*, 97; "Frankreichs Biereinfuhr in Deutschland," *Der Schwäbische Bierbrauer* 11, no. 16 (April 16, 1882): 123–124.

31. Jeffrey M. Pilcher, "National Beer in a Global Age: Technology, Taste, and Mobility, 1880–1914," *Quaderni storici* 151, no. 1 (April 2016): 51–70; R. G. Wilson, "The Changing Taste for Beer in Victorian Britain," in *The Dynamics of the International Brewing Industry since 1800*, ed. R. G. Wilson and T. R. Gourvish (London: Routledge, 1998), 93–104; Ranjit S. Dinghe, "A Taste for Temperance: How American Beer Got to Be So Bland," *Business History* 58, no. 2 (2015): 1–32; Malcolm Purinton, "Empire in a Bottle: Commerce, Culture and the Consumption of the Pilsner Beer in the British Empire, 1870–1914" (PhD diss., Northeastern University, 2016).

32. "Ueber die Gründung neuer Aktienbrauereien," *Wochenschrift für Brauerei* 5, no. 25 (June 22, 1888): 497.

33. "Die Dampfkessel-Explosionen im Deutschen Reiche während des Jahres 1892," *Wochenschrift für Brauerei* 10, no. 38 (September 22, 1893): 1018.

34. On an early pioneer, see Michaela Knorr, "Christel Goslich," in *125 Jahre Versuchs- und Lehranstalt für Brauerei in Berlin e.V.* (Berlin: VLB, 2008), 53.

35. "Rückgang des Konsums des Wiener Bieres," *Zeitschrift für das Gesamte Brauwesen* 22, no. 1 (1900): 15.

36. Karl Dubský, "Uiber Geschmacksreinheit der Biere," *Der Böhmische Bierbrauer* 26, no. 19 (October 1, 1899): 565.

37. Karl Dubský, "Ernste und heitere Betrachtungen eines Bierphilosophen," *Der Böhmische Bierbrauer* 26, no. 16 (August 15, 1899): 475.

38. Dave Infante, "Craft Beer's Moral High Ground Doesn't Apply to Its Workers," Splinter, accessed May 21, 2018, https://splinternews.com/craft-beer-s-moral-high-ground-doesnt-apply-to-its-work-1826080180.

5

Modern Food as Distributed Food

THE THIN RIPE LINE
WATERMELONS, PUSHCARTS, DISTRIBUTION, AND DECAY

William Thomas Okie

Modern food is Burkina Faso green beans on Paris shelves in February, California strawberries in Toronto in March, Chilean grapes on Chicago tables in April, Georgia peaches in Boston in May.[1] Modern food is the roughly 2,000 miles, from five foreign countries, that your meal has traveled to reach your plate; it is the fact that farmers' markets, pick-your-own operations, and other direct sales still represent less than one-third of 1 percent of fresh food consumed.[2] Modern food, in terms of fresh fruits and vegetables, is distributed food.

But unlike production and consumption, the process of distribution is dull. It calls to mind neither the virtuous farmer with her straw hat, loam-covered boots, and hands full of berries, nor the epicurean privileges of the consumer picking out just the right artisanal olives. The organic and local food movements have done a great deal to call attention to the social and ecological dimensions of our food system, but in their most prominent manifestations have done little to explore the distribution of food. The "supermarket pastoral," the "Who's Your Farmer?" bumper stickers, even the phrase "farm-to-table" itself—all sustain the fiction of intimacy between producer and consumer, as well as the conceit that such intimacy could overcome the inequities and injustices of the modern food system. And all ignore the intricate world contained in that tiny preposition "to."

It's easy to ignore distribution now, if you live in the world of super-markets and factory farms of the twenty-first century. But in the years around the turn of the twentieth century, as the modern food system we know and love or hate today was taking shape, the work of distribution was a matter of legs and lungs. The problems associated with food distribution assaulted the senses and occupied the attention of everyone from impoverished housewives to middle-class reformers, from struggling farmers to government officials, from fresh-off-the-boat immigrants jammed together in tenements to plant pathologists housed in universities. These folks all argued over distribution. They developed new processes to distribute a wider range of foods to farther-flung places. And they understood all too well that to grapple with food distribution was also to grapple with decay. Instead of emphasizing the role of technological fixes like refrigeration and rapid transit, this essay foregrounds the biological forces that shaped the distribution of modern food. This is a reminder that "ripe" is just another word for "almost rotten," that "fresh" can be a synonym for "not yet spoiled." This is a story about decay—about fearing it, fighting it, and living with it.

The world of distribution is one of refrigerated tractor trailers, vast impersonal warehouses, container ships, railroad terminals, commission merchants, and government inspectors; it is also a world of fungi, bacteria, and putrefaction. One might appreciate the importance of all those things, but probably in the same way one appreciates the importance of hydraulic pressure in car brakes and the water resistance of asphalt shingles. Perhaps a death scene would be more intriguing?

It's July 19, 1895, and we are with a *New York Times* reporter on the Hudson River pier of a prominent steamship company in lower Manhattan.[3] Longshoremen emerge from the hold of the ship with load after load of watermelons from Georgia plantations. New York City is the center of the nation's produce trade, and the steamer's load of 70,000 watermelons is just one in a long litany of produce from around the world newly available in the markets of New York City: oranges from Sicily; black cherries, plums, currants, and huckleberries from California; and peaches from Georgia. The global shipping industry is circulating new foodstuffs to new places, much like the wheat traveling from California to England in Thomas Finger's contribution in this collection (chapter 1).

But something is rotten here. An "almost unbearable" stench fills our nostrils. The wooden planks we are standing on appear to have been marinated in putrefying watermelon pulp, and the place reeks with a moist, sickly alcoholic scent. A foul pink slime coats the tracks of the carts taking the melons away. The longshoremen gamely stack the melons in pyramids along the water's edge as long as the yellowing rinds hold together, but many fall apart—and not with that familiar crisp thunk but a dispiriting pulpy swish.

Approximately half of the melons emerging from the ship, some 30,000 in all, are yellowed and decayed, and almost no one seems to want them. Certainly not the produce merchants who ordered them from Georgia and have the option of inspection and refusal upon arrival. Nor the longshoremen, who look askance at the decaying fruit, nor even the street children, though they are everywhere in 1890s Manhattan and are always hungry.

Yet the steamship company appears to be doing a brisk business, selling whatever is not obviously rotten at "practically any price." Italian fruit dealers are "out in force" with what the New York Times correspondent describes as "ramshackle carts" pulled by "decrepit horses," "merrily" going away with thousands of decaying melons to be distributed to the unsuspecting poor of New York and Jersey City. And of course, 30,000 rotten watermelons constitute not just shady business but a threat to public health. "The very thought of the condition of these melons," a produce merchant tells us, "is enough to give a man cholera morbus." And yet there are no health inspectors in sight, no "sanitation squad" who might put things right by seizing, condemning, and dumping the putrid load into the river.

We can learn a lot about the origins of modern food from one gruesome scene.

For one thing, watermelons were time bombs of decay. Widely grown and much beloved across the nation in the nineteenth century, watermelon (*Citrullus lanatus*) as a commercial crop was new in the late nineteenth century and was mostly a southern phenomenon. By the 1890s the industry covered more than 50,000 acres; in 1895 Georgia growers alone shipped some 10 million melons to northern markets.[4] That July the *New York Times* was full of reports of ripe melons at by-the-slice stands where "crowds of people . . . engaged in devouring them by the dozen."[5] Scientist Hugh Starnes described watermelons as "by far the most sensitive of all perishable horticultural products" and recommended extreme care in harvesting, handling, shipment, and especially market selection. "When over-ripe," Starnes wrote, "there begins to creep over the surface of the melon . . . an unmistakable, faint, sickly-white tinge, turning later to a bilious yellow."[6]

And there it is: the rot and decay. Few of us think about it as much as our early twentieth-century counterparts did. US Department of Agriculture (USDA) pomologist G. Harold Powell, for instance, wrote in 1906 that "No commodity is more likely to deteriorate in transit than the fruit crop" especially during "hot, moist shipping seasons."[7] The "causes of losses" Powell described were extensive and varied. A scab called bitter-rot grew "luxuriantly in warm weather" on apples and then provided a foothold for blue and pink mold; blue molds likewise afflicted pears, lemons, and oranges. *Botrytus* attacked strawberries, especially when the fruit's "vital processes are at low ebb" or when the skin was broken. Contact injuries like "severe pressing," rubbing, or loose packing, careless handling in the orchard, plantation, and packing house could be the culprits. "Mechanical injuries," whether caused by insect punctures, stems of other fruits, fingernails, windstorms, or excessively rapid fruit growth, also accelerated decay.[8] In short, the path to market was beset with perils, at least for the humans who wished to profit from the fruit. The fruit didn't seem to care much.

Indeed, as Powell acknowledged, the "causes of losses" weren't confined to injuries done to fruit by humans, fungi, machines, or insects, but included the inexorable fact that "the fruit ripens rapidly as soon as it is picked." Fruit seemed to do everything it could to ripen and rot, using its moisture and sugars to tempt its animal consumers to spread its seeds more widely.[9] The rotting of the fruit might be thought of as the distribution system of the plant. Humans could influence, slow, and bound what Powell called this "behavior of fruit" but they could not stop it entirely. Human intervention interrupted the watermelon plant's distribution system because precious few of the seeds left over from melons consumed or trashed in New York City would have much of a chance to germinate and reproduce. Or perhaps it's more accurate to say that humans, using the technologies of telegraphs, steamships, railroads, pushcarts, and plant breeding, complicated the watermelon distribution system for their own ends—which had the effect not of removing humans *from* "nature" but rather, as historian Timothy LeCain put it, embedding them "ever more deeply within its often-unpredictable powers."[10]

Few of "nature's powers" are as unpredictable or underestimated as fungi. As watermelon production expanded and intensified in the South, these organisms expanded and intensified their own use of the plants for their own ends—ends that included, of course, distribution. Anthracnose fungi (*Colletotrichum lagenarium*) sank their threadlike roots into the epidermal walls of leaf and melon alike and had the "peculiar ability" to lay dormant until a warm, wet microclimate was available, such as that in a truck or railroad car headed north, and then rapidly transform the smooth green surface into a moonscape of sunken spots and pink spores. Similarly, stem-end rot (*Diplodia*) typically showed up in transit, where it could turn a carload of market-ready fruit into a festering mass of slimy brown skin and black mold in a matter of days.[11]

Because of all of this under-the-surface biological activity, distribution of produce as food was a delicate task. Watermelons might grow well in the vast fields and long hot summers of the South; they might earn fancy prices from overheated city-dwellers. But unlike some perishable goods, like the ones Anna Zeide describes in chapter 10, watermelons could not be preserved. Their field-to-table transit had to be rapid and required infrastructure like the steamer, the pier, the markets, and the wagons and carts. It also required energy—the climate energy of ice, the fossil fuel energy of coal-fired trains, the biological energy of the longshoremen and horses, the brains of the Italian produce sellers, even the opportunism of the street children.

New York City's marketing facilities of steamship piers, railroad terminals, and produce markets were under immense strain at the turn of the twentieth century. The city's population had increased more than sixfold in fifty years, and all those people had to live and eat and defecate in roughly the same amount of space.[12] One USDA official pictured the problem as a matter of shapes: "We have piled our people up in cities twenty stories high in great communities such as have never existed before . . . and those communities produce not one thing they eat." To feed this vertical population, the city's horizontal supply network stretched further and further—like a root system expanding toward the horizon as the tree reaches toward the sky. In the nineteenth century local hinterlands had supplied most of the city's food needs; indeed, cities fed themselves to a great extent with kitchen gardens along sidewalks and livestock in the streets. But by the 1920s the average produce item traveled 1,500 miles to the great city, prompting one pamphleteer to boast: "ALL THE WORLD FEEDS NEW-YORK."[13] If all the world fed New York, much of the world's produce would have to come through a few blocks along the Hudson River in Lower Manhattan. This bottleneck situation worried city officials and progressive reformers, who feared, whether for humanitarian or political reasons, a municipal famine, an actual shortage of food for the city's working class. It also created an unprecedented opportunity for street vendors, most of whom were recent Italian and Jewish immigrants who plied their trade with pushcarts.

A reliable distribution process eluded growers, marketers, and consumers alike. Farmers organized exchanges and cooperatives, bargained with commission merchants, and sued railroad companies and steamship lines.[14] Consumers staged food riots, called for regulation, and hungrily adopted market advice from newspapers, radio programs, and books.[15] But in New York City, distribution reform efforts were marked by an obsession with the new process of street vending, the very last link in the chain of distribution, and in particular with the human-powered technology of the pushcart.

Street vending had exploded in New York City in the 1860s and 1870s. Peddlers trolled the city's burgeoning wholesale markets, seeking bargains and overstocks, and then roved the city, singing out their wares through labyrinthine neighborhoods, gripping the handles of two-wheeled carts, propping the carts on sticks to make their sales. As the twentieth century opened, the number of pushcart operators trebled in just four years.[16] In 1906, 97 percent of those licenses belonged to immigrants.[17] They were especially important for Jewish and Italian communities on the Lower East Side, where pushcarts gathered in stationary markets, looking for all the world like the open-air village markets that many immigrants had known in Europe.[18] In the early twentieth century a number of grocers had simply given up selling perishables because the work required was too strenuous. Peter Peckich, who owned a grocery in the Bronx, sublet his storefront to an Italian who, when he wanted to sell berries, would leave at 10:00 p.m. on his horse for Washington Market, spend the night at the market, purchase his fruit first thing in the morning, and bring it back to the store by 10:00 a.m.[19] In a similar way, pushcart peddlers filled a gap in the city's fresh food distribution system. They may have sold the cheapest goods in individual volumes too small to count, but they performed two crucial functions: relieving market gluts by buying surplus produce and providing fresh food for the residents of tiny apartments in increasingly dense neighborhoods (see figures 5.1 and 5.2 for typical pushcart scenes in the 1910s).

Pushcarts also attracted a tremendous amount of attention from progressive reformers and city officials. This concern is obvious in the 1895 watermelon story, in which "the Italians" play the villains, gathering up "unfit" produce and merrily profiting at the expense of the poor uneducated

5.1   Peddler, East Side, ca. 1915. (Bain News Service, George Grantham Bain Collection, LC-B2-4004-14 [P&P] LOT 10915, Library of Congress Prints and Photographs Division, Washington, DC, http://hdl.loc.gov/loc.pnp/ggbain.22928.)

5.2 Pushcart market, New York, ca. 1910. (Bain News Service, George Grantham Bain
Collection, LC-B2-3481-9 [P&P], Library of Congress Prints and Photographs Division,
Washington, DC, http://hdl.loc.gov/loc.pnp/ggbain.19115.)

population of the city. But it was a common theme of reportage in the
1880s and 1890s. In 1893, for instance, a *New York Times* reporter took a
stroll down Hester and Essex Streets and enumerated the offenses of the
mostly Eastern European neighborhood: a couple of Romanians selling
"putrid fish" and slinging fish scales and slime on the "black as tar" Polish
bread the next stand over; a cart of meat and fowl swarming with "great
blue flies" and "already alive" with maggots; a "slatternly" and scantily clad
young woman selling cheese the reporter described as a "reeking mass of
rottenness and alive with worms"; fruit stands piled high with half-decayed
fruit; skin disease and filth and vermin and trash everywhere; and streets
"impassable" for anything larger than a pushcart. The neighborhood, the
reporter concluded, "perhaps the filthiest place on the Western Continent,"
was inimical to Christian virtue and an open invitation to cholera.[20] A couple of weeks later, thirty-one pushcart peddlers were arrested for obstruct-
ing the streets and selling decayed fruit, and Hester Street got a thorough

cleaning: the police routed out the pushcarts and wagons, the street cleaning department carted away the trash and swept the pavement, the fire department flushed the street with water, and the "sanitary squad" then paraded down the street sniffing for any leftover garbage.[21]

Pushcart peddlers were thus quite visible and, as recent immigrants, quite vulnerable. Along with broader concern about the problem of distribution, this vulnerability and visibility created what became known as "the push-cart evil," with attempts to regulate pushcarts taking place in every decade of the twentieth century.[22] Pushcarts attracted special mayoral commissions and studies in 1906, 1913, 1915, and 1917, the last of which argued that in the decade since the 1906 report, "the pushcart problem has not changed materially."[23] In the 1930s Mayor Fiorello LaGuardia spearheaded a bizarrely single-minded campaign to enclose all pushcarts in "suitable market buildings."[24] At the opening of one of his new indoor markets, LaGuardia looked around at his audience of vendors wearing "crisp white coats" and declared: "I found you pushcart peddlers. . . . I have made you MERCHANTS!"[25]

Like many in the Progressive Era's educated class, reformers displayed a fervent faith in scientific data and were willing to put in a great deal of effort to understand and document this "pushcart problem." To take just one example, the 1906 pushcart commission partnered with the police department to complete what we might call the Great Pushcart Census. Hundreds of police officers descended into the streets over a period of several weeks armed with stacks of survey cards, street maps with precise routes outlined in red ink, and, supposedly, "friendliness and respect" for the pushcart peddlers who were their targets.[26] Each patrolman interviewed between ten and fifteen peddlers to discover the name, nationality, and years of residency of the peddler; whether the peddler owned or rented the pushcart; how much he paid for it; and from whom he purchased his goods. Examining the food, the officer thought of his own children: was this cucumber something he would serve willingly to them? If so, he marked its condition as "good"; if not, he checked "fair," "bad," or "injurious to health"; if the last of these, he gave his assessment of why it was bad, by writing in, for instance, "Decayed." By the end of the summer, police had accounted for 4,515 pushcarts in Brooklyn and

Manhattan. And then the clerical side of the commission went to work, tallying, mapping, and arranging the results of the great pushcart census into ten maps, thirty tables, and more than ninety pages of narrative illustrated with twenty-nine photographs and three diagrams. Pushcart peddlers, the commission charged, had essentially turned the streets of lower Manhattan into open-air markets, causing "great inconveniences" which, if unaddressed, would soon yield "serious evils."

Pushcart peddlers responded by portraying themselves as honest businessmen with their eyes on the twin prizes of citizenship and a more stable occupation, and in the process providing a public utility, preventing food waste and allowing for distribution of perishable foods into the interior of the city. One Italian representative claimed "more than one million clients . . . who daily go to the push-carts," collectively spending $28,000 a day. In part, the peddlers claimed to agree with the need for more regulation to distribute the carts more evenly about the city or require licenses with photographs. But many vehemently opposed the creation of central markets: "To congregate the peddlers in one or ten places in all the city," the Italian spokesman explained, "would mean destruction for the class, and starvation for 14,000 families."

The resulting report was ambivalent. In a manner strikingly prescient of our twenty-first-century preoccupation with food trucks, it acknowledged that pushcarts added "materially to the picturesqueness of the city's streets," imparted an "air of foreign life which is so interesting to the traveler," and lent "an element of gaity and charm to the scene which is otherwise lacking." Yet "the practical disadvantages from the undue congestion of peddlers in certain localities are so great as to lead to a demand in many quarters for the entire abolition of this industry." Despite these disadvantages, the report ultimately recommended relatively modest reforms: changing the licensing rules to allow fresh-off-the-boat immigrants to peddle and enforcing a more even distribution of pushcarts on certain crowded streets, but otherwise leaving the system as it was. The pair of photographs that opened the report encapsulated the sort of reform they hoped to see: traversable streets that retained an air of the picturesque (see figure 5.3). As historian Daniel Bluestone notes, the pushcart reformers, in 1906 as well as in subsequent reports, articulated a

5.3   Orchard Street, as it is and new proposal, ca. 1906. (Science, Industry and Business Library: General Collection, The New York Public Library, New York Public Library Digital Collections, accessed February 2, 2018. http://digitalcollections.nypl.org/items/9e9c2.)

"narrow view of the street as a traffic artery," no longer available for marketing or for other uses such as "political activity, gregarious socializing, and popular amusements."[27] The primary purpose of a street, according to the report and others like it, was movement.

One of the surprising findings of the report, in view of the complaints about the foul pushcart markets of the 1890s, is that pushcart food was remarkably fresh. Only 3 percent of pushcarts sold unfit food, and only 0.25 percent of the food was "more or less decayed." An average cart had one hundred pounds of food. On any given cart, only one apple or cucumber (about four ounces) was likely to be rotten. Compared to stores (13 percent unfit) and stands (36 percent unfit) pushcarts sold some of the safest produce in the city.[28] Pushcart operators were hot-potato experts.

As a means of distribution, then, pushcarts may have been disorganized, but they were also direct and they reduced food decay. Perhaps they were more direct *because* they were more disorganized. For food to get from the pier on the Hudson River to the distant boroughs, someone had to navigate the streets and cart it to the consumer, with their own strength or the strength of their horses. The reformers' dream was to transform chaos into order without losing the distributive efficiency that seemed to be integral to the chaos. In the midst of the pushcart reform era, some cautioned that this dream might well be impossible. "I do not know of any human foresight or power that can establish a Market," commission merchant William H. Behrenberg testified in 1913. "It is a matter of growth and of a great number of circumstances which no man can foresee."[29] Or as Manhattan borough president Marcus Marks put it in 1915, "Markets are not built; they grow."[30] The social politics governing civic reform also structured the food distribution system itself. The pushcart reformers sought to turn unseemly foreign peddlers into law-abiding American merchants, and to do so by attacking the fundamental problem that gave rise to pushcarts in the first place: rot and decay.

If the 1895 watermelon fiasco and subsequent pushcart debates illustrated the larger turn-of-the-century problem of distribution, it also showed that the problem of distribution was inseparable from the much more ancient problem of decay. As Marks and Behrenberg unwittingly testified, a market is not just *like* a natural thing, it is *actually* natural, inextricable from the material world we inhabit. We speak now of markets as places only in an abstract sense, the forces of supply and demand meeting in a meta-abstraction that we call "the economy."[31] Even in the twenty-first century markets are material, down to the sand mines that supply the silicon that comprises the microchips that house the human-authored algorithms that power much trading.[32] But in the 1890s and 1900s, the biological materiality of marketing and distribution was especially hard to avoid.

The weather mattered, for instance. When Georgia Experiment Station scientist Hugh Starnes advised southern watermelon growers on "gathering and marketing" in 1897, his most important piece of advice was to watch the weather. "On a chilly day in summer," he wrote, even "an insignificant shipment of a few carloads will glut beyond hope a market that on a hot, dry day would swallow them by the trainload and eagerly call for more."[33] Similarly, in a typical August 1926 report from two New York commission merchants, they reported on the weather in the markets and its effect on the condition of the product and the demand, writing, "Weather is just hot enough to make buyers here careful as to the quantity of peaches that they may purchase for fear that the fruit will not hold up well."[34]

Watermelons ripened, fungi multiplied, and humans hungered. What at first glance seems to be a routine journey of a watermelon from a southern field to a northern market on a hot July day is, on closer inspection, a complicated tangle of competing and cooperating instinct, desires, and breeding. And the overriding instinct, of course, for all the organisms involved was the perpetuation of the species, the continuation of life. To distribute fresh fruits and vegetables was to engage in an act of death defiance.

All of which brings us back to the fear of municipal famine that fed the market reform fervor and led to the century of pushcart reports. The problem of markets was the problem of sustaining human *life* in a large city. And from the perspective of a late nineteenth-century New Yorker, the produce merchant's joke in 1895 about the rotting melons causing "cholera morbus" was not really very funny. Cholera was one of the most feared diseases of the nineteenth century, and the summer of 1892 had seen severe epidemics in the Middle East, continental Europe, and especially Russia, where an estimated 620,000 cases resulted in over 300,000 deaths.[35] Cholera, its chroniclers say, was "more than a killer; it degraded its victims"; it was "a vulgar and demeaning disease."[36] Death by cholera begins with vomiting and diarrhea so severe that it evacuates pieces of the intestinal lining itself, giving the stools a distinctive "rice water" appearance, followed in relatively short order by dehydration, seizures, incoherence, and coma.[37] Death frightened; death by cholera terrified. And although in retrospect we might surmise that cholera was much more likely to be in the water that washed the fruit than in the fruit itself, at the time there was widespread agreement that disease went hand in hand with filth and decay.[38] There was something cosmically disturbing about the idea that in seeking to sustain the lives of your children by purchasing some fresh fruit, you could actually destroy them by exposing them to disease. Perhaps it was for this reason that when city officials commenced the great pushcart census of 1906, they instructed police officers to ask whether or not they would willingly feed the food they inspected *to their own children*—as if to make it clear that pushcart reform was not merely about order but about reproduction, the biological sustainability of the next generation of human society in that great city.

Food scares remain familiar. To take just one recent year, 2018 saw multistate outbreaks of *Salmonella* in tahini, cut melons, eggs, frozen and dried coconut, raw sprouts, pasta salad, and Kelloggs Honey Smacks cereal; *Escherichia coli* in romaine lettuce and ground beef; *Listeria* in deli ham and ready-to-eat pork patty rolls; *Coclospora* in McDonald's salads and Del Monte vegetable trays; and *Vibrio* in fresh crab meat.[39] While regulation and investigation are much more extensive than they were a hundred years ago, food distribution and therefore the chance for cross-contamination is even more complex. In the first *E. coli* O157:H7 romaine outbreak of 2018, in which 210 people fell ill and 5 died, the U.S. Food and Drug Administration determined that the common edible denominator was chopped romaine lettuce in restaurants (and at least one prison) in thirty-six states—all from contaminated irrigation canals in Yuma County, Arizona.[40]

Decay is a historical force. In the case at hand, decay of fruit created an economic niche for peddlers because the fruit wholesalers and merchants had to sell the fruit quickly before the sugars that make it marketable succumbed to fungi that would render it worse than worthless. Decay also shaped demand for the peddlers' services, for of course the act of eating prevented the perishing of the peddlers' chief customers, the poor residents of New York City's crowded tenements. And it was fear of infra-structural and social decay that led reformers of the "pushcart problem" to study it and seek to transform it through new processes of regulation and enforcement. New York City pushcart peddlers, with their biologi-cally powered marketing technology, filled not just an economic niche but completed a crucial biological link between rapidly ripening fruit and their customers, with their biological need to eat.

The line between ripe and rotten is thin. Distribution may be at the center of the food system, but has been mostly peripheral to the contem-porary food movement. Packing houses, refrigerator coils, tractor trailers, and supermarket dumpsters are decidedly not the faces of the supermarket pastoral. But perhaps they should be. Perhaps, in other words, we need to catch up with our early twentieth-century counterparts, to go beyond remembering the farm to also remember the bodies and fungi and matter and life and death that make up the world of distribution, where the cen-tral lines of action form around the facts of putrefaction.

NOTES

1. Susanne Freidberg, *French Beans and Food Scares: Culture and Commerce in an Anxious Age* (New York: Oxford University Press, 2004); Miriam J. Wells, *Strawberry Fields: Politics, Class, and Work in California Agriculture* (Ithaca, NY: Cornell Univer-sity Press, 1996); Heidi Tinsman, *Buying into the Regime: Grapes and Consumption in Cold War Chile and the United States* (Durham, NC: Duke University Press, 2014); William Thomas Okie, *The Georgia Peach: Culture, Agriculture, and Environment in the American South* (New York: Cambridge University Press, 2016).

2. Natural Resources Defense Council, "Food Miles: How Far Your Food Travels Has Serious Consequences for Your Health and the Climate," *Health Facts*, November 2007, https://food-hub.org/files/resources/Food%20Miles.pdf; Leopold Center for Sus-tainable Agriculture, "How Far Do Your Fruit and Vegetables Travel?," *Leopold Letter* 14, no. 1 (Spring 2002): 9, https://lib.dr.iastate.edu/cgi/viewcontent.cgi?article=1021 &context=leopold_letter; Worldwatch Institute, "Globetrotting Food Will Travel Farther Than Ever This Thanksgiving," accessed June 8, 2018, http://www.worldwatch .org/globetrotting-food-will-travel-farther-ever-thanksgiving. According to the USDA, the number of farmers' markets in the United States nearly quadrupled between 1994 and 2016, from 1,755 to 8,669. See https://www.ams.usda.gov/sites/default/files/me dia/National%20Count%20of%20Operating%20Farmers%20Markets%201994-2016

.jpg. As of the last agricultural census in 2012, 144,530 farms sold fresh farm products directly to consumers for a combined total of $1.3 billion—an 8 percent increase since 2007. See United States Census of Agriculture, "Farmers Marketing," *2012 Census of Agriculture Highlights* (August 2014), https://www.agcensus.usda.gov/Publications /2012/Online_Resources/Highlights/Farmers_Marketing/Highlights_Farmers_Market ing.pdf.

3. "Decayed Georgia Melons: A Cargo of 30,000 Sold by a Steamship Company," *New York Times*, July 20, 1895.

4. "Where Watermelons Come From," *New York Times*, July 19, 1871; Lee J. Vance, "Millions of Melons," *Harper's Weekly*, September 26, 1896.

5. "In the Retail Markets: Ripe Melons and Other Fruits in Good Supply," *New York Times*, July 29, 1883.

6. Hugh N. Starnes, *Watermelons*, Georgia Experiment Station Bulletin 38 (Atlanta: Georgia Experiment Station, 1897), 77.

7. G. Harold Powell, "The Handling of Fruit for Transportation," in *Yearbook of the United States Department of Agriculture 1905* (Washington, DC: Government Printing Office, 1906), 350, http://archive.org/details/yoa1905.

8. Powell, "The Handling of Fruit for Transportation," 351–354.

9. Powell, "The Handling of Fruit for Transportation," 361, 356.

10. Timothy J. LeCain, *The Matter of History: How Things Create the Past* (Cambridge: Cambridge University Press, 2017), 66.

11. W. A. Orton, *Watermelon Diseases*, Farmers' Bulletin 821 (Washington, DC: US Department of Agriculture, 1917), http://archive.org/details/CAT87202542.

12. New York's population increased from 515,547 in 1850 to 3,437,202 in 1900. "Total and Foreign-Born Population, New York City, 1790–2000," NYC.gov, accessed July 9, 2018, https://www1.nyc.gov/assets/planning/download/pdf/data-maps/nyc -population/historical-population/1790-2000_nyc_total_foreign_birth_pdf.

13. Mary Pennington, "The Proper Handling of Foodstuffs from Farm to Market," in New York City Market Commission, *Report of the Mayor's Market Commission of New York City* (December 1913), 252; Marc Linder and Lawrence Zacharias, *Of Cabbages and Kings County: Agriculture and the Formation of Modern Brooklyn* (Iowa City: University of Iowa Press, 1999); Catherine McNeur, *Taming Manhattan: Environmental Battles in the Antebellum City* (Cambridge, MA: Harvard University Press, 2014). "ALL THE WORLD FEEDS NEW-YORK," from Walter P. Hedden, *Produce Terminal Requirements in the New York Area* (New York: Port of New York Authority and the Bureau of Agricultural Economics, 1925), 4.

14. Okie, *The Georgia Peach*, chap. 5; Victoria Saker Woeste, *The Farmer's Benevolent Trust: Law and Agricultural Cooperation in Industrial America, 1865–1945* (Chapel Hill: University of North Carolina Press, 1998); Douglas Cazaux Sackman, *Orange Empire: California and the Fruits of Eden* (Berkeley: University of California Press, 2005).

15. Hasia R. Diner, *Hungering for America: Italian, Irish, and Jewish Foodways in the Age of Migration* (Cambridge, MA: Harvard University Press, 2001); Dana Frank,

"Housewives, Socialists, and the Politics of Food: The 1917 New York Cost-of-Living Protests," *Feminist Studies* 11, no. 2 (July 1, 1985): 255–285; Meg Jacobs, *Pocketbook Politics: Economic Citizenship in Twentieth-Century America* (Princeton, NJ: Princeton University Press, 2005); Tracey Deutsch, *Building a Housewife's Paradise: Gender, Politics, and American Grocery Stores in the Twentieth Century* (Chapel Hill: University of North Carolina Press, 2010).

16. In 1900, 2,073 licenses had been issued; in 1904, 6,747. *Report of the Mayor's Push-Cart Commission* (New York: City of New York, 1906), 11.

17. *Report of the Mayor's Push-Cart Commission*, 90.

18. Suzanne Wasserman, "Hawkers and Gawkers: Peddlers and Markets in New York City," in *Gastropolis: Food and New York City* (New York: Columbia University Press, 2010), 155.

19. Testimony of Peter Peckich, in New York City Market Commission, *Report of the Mayor's Market Commission of New York City*, 244–245.

20. "East Side Street Vendors: Their Push Carts Obstruct Many Streets," *New York Times*, July 30, 1893, sec. 17.

21. Melanie A. Kiechle, *Smell Detectives: An Olfactory History of Nineteenth-Century Urban America* (Seattle: University of Washington Press, 2017).

22. City of New York Department of Business Services, "New York In Transition: Itinerant Peddlers and Vendors Everywhere" (November 1992), i.

23. New York City Market Commission, *Report of the Mayor's Market Commission of New York City*; Marcus M. Marks, *Reports on Market System for New York City and on Open Markets Established in Manhattan* (New York: Borough of Manhattan, City of New York, 1915), x-devonthink-item://8DE1D014-7DB2-498D-9BB3-729E019536B7; Leonard M. Wallstein, "The Pushcart Problem in New York City: Progress toward Its Solution and Need for Prohibition of Pushcart Peddling in the Vicinity of the Williamsburg and Manhattan Bridge Markets" (New York: Commissioner of Accounts, 1917), 4; Daniel Bluestone, "The 'Pushcart Evil,'" in *The Landscape of Modernity: New York City, 1900–1940*, ed. David Ward and Oliver Zunz (New York: Russell Sage Foundation, 1992), 299–300, 302–305; Wasserman, "Hawkers and Gawkers," 157–158, 160–164.

24. Benjamin Koenigsberg, "Essex Street Market Location," Commissioner William Fellowes Morgan to *East Side Chamber News*, December 9, 1936, 8, cited in Wasserman, "Hawkers and Gawkers," 160.

25. Newbold Morris, *Let the Chips Fall: My Battle against Corruption* (New York: Appleton-Century-Crofts, 1955), 119–120, cited in Wasserman, "Hawkers and Gawkers," 163.

26. Quotes in this and the next three paragraphs are drawn from New York City Push-Cart Commission, *Report of the Mayor's Push-Cart Commission* (New York: City of New York, 1906), 121, 122.

27. Bluestone, "The 'Pushcart Evil,'" 287.

28. New York City Push-Cart Commission, *Report of the Mayor's Push-Cart Commission*, 57–58.

29. New York City Department of Public Works, *Preliminary Studies for a Fresh Fruit and Vegetable Wholesale Produce Market, New York City*, vol. 1, *Factual Data* (New York: Department of Public Works, 1946), 26.

30. Marks, *Reports on Market System for New York City*, 37.

31. See Timothy Mitchell, "Fixing the Economy," *Cultural Studies* 12, no. 1 (January 1998): 82–101; William Cronon, *Nature's Metropolis: Chicago and the Great West* (New York: W. W. Norton, 1991).

32. See LeCain, *The Matter of History*, 133.

33. Starnes, *Watermelons*, 77.

34. Smith and Holden to Lamartine G. Hardman, August 4, 1926, Lamartine G. Hardman Sr. Papers, Series II, Box 74, Folder 1, Richard B. Russell Library, University of Georgia.

35. Howard Markel, *Quarantine! East European Jewish Immigrants and the New York City Epidemics of 1892* (Baltimore: Johns Hopkins University Press, 1999), 86.

36. Michael Zeheter, *Epidemics, Empire, and Environments: Cholera in Madras and Quebec City, 1818–1910* (Pittsburgh: University of Pittsburgh Press, 2016), 7; Richard J. Evans, *Death in Hamburg: Society and Politics in the Cholera Years, 1830–1910* (Oxford: Clarendon Press, 1987), 230.

37. Markel, *Quarantine!*, 87.

38. "Sickness Caused by Decayed Fruit (from the Chicago Times-Herald)," *New York Times*, September 29, 1895, sec. 25.

39. Centers for Disease Control and Prevention, "List of Selected Multistate Foodborne Outbreak Investigations," accessed February 8, 2019, https://www.cdc.gov/foodsafety/outbreaks/multistate-outbreaks/outbreaks-list.html.

40. "Multistate Outbreak of *E. coli* 0157:H7, June 28, 2018, https://www.cdc.gov/ecoli/2018/o157h7-04-18/index.html; US Food and Drug Administration, "FDA Investigated Multistate Outbreak of *E. coli* O157:H7 Infections Linked to Romaine Lettuce from Yuma Growing Region," November 1, 2018, https://www.fda.gov/Food/Recalls OutbreaksEmergencies/Outbreaks/ucm604254.htm.

TRUST

6

Modern Food as Suspicious

GILDED SUGAR AND CORN SYRUP'S LONG CON

Benjamin R. Cohen

Corn syrup began as a scam a long time ago. Many people still think of the sweetener that way, especially in its high fructose version. They worry over health. They worry about identity and transparency. In 2018 a columnist in the *New York Times* took it as his prompt that Big Sugar has "spent years trying to trick you." When you sell me corn syrup, are you selling me something legitimate? Given the world-historical prevalence of sweeteners in the marketplace, underneath the concerns is a deeper question asking, are all sugars the same?[1]

A five-year court battle settled in 2015 tried to tackle that last one. The case of *Western Sugar Coop, et al. v. Archer Daniels Midland Co., et al.*, resolved that, legally speaking, all sugars were not the same, that chemical formula alone did not make for authentic similarity. In the lawsuit the court ruled that the Corn Refiners Association (CRA) was guilty of false advertising. The CRA's product was Karo Syrup, which is plain ol' corn syrup. They labeled Karo bottles "sugar." The legal decision said, no, you can't do that anymore; that would be deceptive. *Saturday Night Live* had parodied the CRA's ads, mocking the idea that refined syrups—that's Karo—were the same as field-grown sugars like cane. Even if *SNL*'s viewers were okay with corn syrup or its high fructose version, they had to know those weren't the same thing as cane. Budweiser later took suspicion of corn syrup as a given in a 2019 ad campaign. With nary a backstory, they mocked their beer competitors for using the syrup as a fermenting agent instead of presumably natural and healthier ingredients (Budweiser uses rice). It was self-evident to them that corn syrup is lesser.[2]

As it is, corn syrup has been a troublemaker from its earliest days in the nineteenth century. Long before the historical periods that we imagine as important precursors, corn syrup got its start in the 1860s and 1870s. At the time it had a number of names. Some people called it grape sugar, because it had first been made from grapes. Or they called it glucose, a chemical term devised during a time when devising chemical terms was a new pastime, as Lisa Haushofer describes in chapter 12. Today glucose is blood sugar, a carbohydrate in your body. Not so before the twentieth century. Its makers sold it to dress up table food, which led to yet another nickname, table syrup, using it as a substitute for other sugars like cane and beet. And these glucose/grape sugar/table syrup makers confused people who weren't sure what the product was or why it was there.

We've always doubted corn syrup. Suspicion over its identity and validity has embodied suspicion over the modern food marketplace since its early years. If we want to do anything about that suspicion, that sense that *something is not quite right in our food system*, we'll need to stop thinking the solution is a new kind of sweetener and find from history that it's the doubt we need to attend to, the trust in processing and refining. This is a story about where that suspicion came from as new products entered the marketplace in ever higher numbers in the late nineteenth century.

There are any number of reasons to mistrust something or someone. Unfamiliarity, confusion, miscommunication, lack of experience with something new, or too much prior experience with sleight of hand. Mistrust of food has existed as long as there has been tradable food. The Bible weighed in on it. Plato wrote about it. It isn't new. As the range of stories in this volume further attest, the latter half of the nineteenth century stepped up those worries by changing what food meant and where it came from with an entirely new era of production and consumption. Corn processing offers a glimpse into a transition from an agrarian world of food and rural livelihood to an industrial one of manufactured commodities. As with canned food (see Anna Zeide's discussion in chapter 10), consumers were skeptical of a new industrial world of manufactured commodities when they were skeptical of corn syrup.

There were specific reasons to mistrust glucose (or grape sugar, or table syrup), from shady language and secrecy to manufacturing processes and health scares. I'll get to them soon. First, though, were the larger contexts that gave meaning to the specific kinds of mistrust. Glucose was produced not just by new factories, but by a thicker world of cultural uncertainty and ambiguity in the sweetener marketplace.

A confidence scheme perpetrated in 1874 by the Grangers' Glucose Company offers a glimpse into the world that birthed glucose. It stood astride a renewed nation not a decade beyond the Civil War, with a tantalizing but opaque key term in each hand—Grangers and glucose. With its tidy four-page business proposal, the company was basically an attempt to separate farmers from their money.

Edward Nieuwland wrote the proposal. He was an immigrant from Holland now lost to the popular annals of history, landing in a busy Manhattan in the early 1870s. His scene was like those in PBS documentaries, the ubiquitous soundtrack with clanking carts and clip-clop horse carriages and hawkers selling wares on the sidewalk. These were the streets full of peddlers gripping the handles of two-wheeled carts Tom Okie describes in chapter 5.

Nieuwland walked down the middle of a city rapidly expanding, bustling, and ripe for schemes. He had big bushy sideburns that people said made him the spitting image of railroad tycoon Cornelius Vanderbilt—stout, important, commanding.[3] Before he came to New York, Nieuwland had been a money broker in Europe. When he got to Gilded Age Manhattan he began a career dotted with attention to currency, labor, capital, and farming, throwing his lot in with a new agrarian movement called the Patrons of Husbandry but known more commonly as the Grange. They were an agricultural advocacy organization, an early lobbying group for rural interests. A grange was an archaic term that had referred to grains and granaries in medieval days but was recently updated in the new American heartland. Its members were Grangers.[4]

And then came the Grangers' Glucose Company. As a model case of confidence scheming, it was a fraud, sneaky and dishonest. It popped up as part of the age of hucksters and charlatans, the Gilded Age, a name drawn from Mark Twain's 1877 coauthored book. The adjective got the emphasis, *gilded*. A gold covering, not gold itself; a shiny surface covering a rotting interior. You couldn't trust what you saw. Those were the streets Nieuwland strolled down. For a story about suspicious food, this was the font of the suspicion. It wasn't clear to farmers or consumers if this novelty syrup was the same thing as sugar or not.

Nieuwland pitched the new opportunity outside the city to Grangers' investors at $50, $100, or $500 a share. He assured potential investors that he "never recommended an investment in which I did not believe myself." He revealed few details about the actual production process except to say that it was "safe and most profitable." It was easy money. While one manufacturer had been a "poor mechanic, now he is the possessor of many millions." The Grangers' Glucose Company could make glucose, this upstart sweetener, better than competitors because Nieuwland had contacts out West where "corn and labor are much cheaper." He could have added that he didn't even need corn—by his time glucose was a sweetener made from any sugar-producing source. An 1881 study found that it could be made from so many sources, including "potatoes, grain, rice, maize, moss, wood, fruits, honey, raisins, etc.," that it didn't even bother to finish the list.[5]

Nieuwland's brokering background helped him conceive of corn as a tool for speculative finance. Rather than a feature of agrarian livelihoods, corn was a commodity defined by its market tradability. The Chicago Board of Trade was founded in the late 1840s as a space to generate traffic for and trade a range of agrarian outputs—corn, grain, cows, and hogs high among them. Glucose, the unknown syrup, fit that world of speculative finance in the 1870s. But was it just gilded sugar?

In the end, it turned out that as an individual scheme, Nieuwland's didn't amount to much. This minor play in the bigger saga of glucose, though, was telling for his awareness of the possible con. He whisked away a legacy of suspicion by showing that the scam was worth trying; it fit into a world where processed foods were neither understood, common, nor trusted. How and why did it get there?

Telescope out from his venture and you find the effort cradled in the nook of the two broader factors that helped set the stage for this fake sugar racket. One was that age of hucksters and con men, the Gilded Age; the other was the general sugar market at the time. At $50 million a year, sugar tariffs were the largest source of federal funds in the late 1800s, as David Singerman tells us in chapter 11. No surprise then that the identity and supposed purity of a sweetener was a big deal. Put the two factors together and you find that the problems of sourcing and identifying a "pure" sugar were lively points of public debate. Recognizing that background about the wider world of sweeteners and diet shows why glucose and corn syrup were trouble from the start.[6]

Sugar was problematic in general, even when it was purportedly "pure," which is to say there were already plenty of reasons consumers found sugar suspect *before* the late nineteenth century.[7] At the time, most sugar came from cane in the tropics or beets in Europe and the American Midwest. The purity of sugarcane from Caribbean and Pacific islands was a long-standing point of deep cultural importance by the time glucose came around.

Cane cultivation was repressive in many ways. Enslaved or conscripted labor formed the basis of the system. "Of the twenty million Africans brought enslaved to the New World," David Singerman writes, "sixteen million owed their perdition to the insatiable Western demand for sugar."[8] Sugar plantation owners justified the needs for that labor with awareness of the environmentally destructive way to cultivate cane. They mowed down forests as plantations moved inland from original coastline plots. Sugar laborers tossed logs in fire boilers nonstop all day. One kilogram of sugar took fifteen kilograms of wood to make. They logged 210,000 tons of wood a year for this.[9] That took a lot of work: those trees didn't cut themselves down and the cane didn't cultivate itself. This was environmental injustice: the repressive labor conditions were tied to destructive environmental practices. They went hand in hand.

Some people acknowledged and acted on that injustice. "Ethical consumption" advocates fighting for fair labor and fair trade today—for coffee, tomatoes, sugar, or against something like "blood" diamonds—echo the responses of sugar activists 200 years ago. In preindustrial times, advocates paved the way for our modern concerns over morality and cultivation by making the case against cane into a case for moral purity. They reasoned that Westerners could lighten the violence of slavery by eliminating sugar from their diets. The author of *An Address to the People of Great Britain* made that point. He argued for boycotts in 1791 with an admirably straightforward point: "If we purchase the commodity we participate in the crime." In abolitionist circles in the antebellum United States, thousands—including Harriet Beecher Stowe and Frederick Douglass—took up a "free produce" movement led by Quakers to encourage abstinence from slave-produced sugar and cotton.[10]

Beyond boycotts, you could also avoid morally impure sugar by turning to beets. Beet sugar, like cane, is sucrose. A good trivia question for friends is to ask which country was the largest producer of sugar in the late 1800s; I doubt they would guess Germany. But along with their French neighbors, the Germans had pioneered sugar beet production. Sugar beets were a more substantial response to cane from the Caribbean and Pacific than growing non-slave-based cane in the continental United States ever was. As historian Kathleen Mapes put it, "The rise of the sugar beet industry over the course of the nineteenth century represents one of the most important, if overlooked, developments in industrial agriculture." In 1884, half of all global sugar came from the sugar beet industry; by 1899 it was 65 percent. As with all matters of food and diet, class and nationalist politics played a role in it all. When capitalists invested with farmers in the Midwest to encourage beet production, they were also advertising a new product that was made in America. They called it a "profitable and patriotic" alternative to cane.[11]

The demand for alternatives to cane cleared the table for upstart sweeteners like glucose. Chemistry played an increasingly valuable role in arbitrating claims of similarity and difference, not just in the Gilded Age but all the way up to Archer Daniels Midland.[12] You develop an appreciation for atoms and molecules when you're trying to figure out what makes one sweetener different from another. Glucose is made of six carbon atoms, twelve hydrogen atoms, and six oxygen atoms ($C_6H_{12}O_6$). So are dextrose and fructose. Sucrose is made of about twice that number ($C_{12}H_{22}O_{11}$). Glucose and fructose are monosaccharides: they are one assemblage of that carbon, hydrogen, and oxygen combination. Sucrose is a disaccharide: it has one part glucose and one part fructose. Our tongues only care so much. They taste reasonably the same, although, really, as sugar historian Deborah Warner explains it, "fructose is somewhat sweeter than sucrose, and glucose is somewhat less sweet."

The part from chemistry you want to hold onto is that it isn't the listed number of elements that matters so much as the way they are arranged. They may differ only slightly in sweetness on the tongue, but our bodies process them differently. It's the process that matters more than the bare identity. You could say this is like the controversy over glucose in the late 1800s. It wasn't the bare fact of a new sweetener derived from potatoes, corn, or whatever. It was the process of how people made it. And where. And for whom.

Today dextrose and glucose refer to the same chemical notation. When they are used differently, it's usually because people are referring to different sources of the sugar. Dextrose is from some production process; glucose is from your body. To make matters more confusing, over the course of the twentieth century the main industry group for fructose took up the process originally built to manufacture glucose. Trying to discern those differences isn't the main point here. In fact, it's the confusion that matters for this story; it's the consistent uncertainty that creates a through line from the 1800s to today. That uncertainty is a feature, not a minor side point, of modern processed foods.

The age of con men and the general sugar market offered two broad explanations for where glucose came from and why it was suspicious and controversial. Those two factors also offered a platform from which the more specific reasons for distrust grew. That gets me back to the workaday reasons across the same late-century era, the shady language, secrecy, manufacturing, and health.

The language of untold profits made easily and swiftly was one key here. Trust me, a guy like Nieuwland was saying, put your faith in me, you'll get rich. Concealment was another tip-off. Early glucose patent holders were sketchy, their secrecy raising suspicion. When one of them, Frederick Goessling, died, it sure did seem like a con when his patents were too vague for anyone else to replicate. Another proprietor "shunned everyone who might discover and reveal the details of his machinery, methods, or business practices." In her study, the historian Deborah Warner found no less than a dozen patent lawsuits at the circuit court level in the early 1880s alone forcing glucose makers to justify their vague claims.[13]

Glucose wasn't usually sold "as is" at the market. Like corn syrup today, it would more likely be an ingredient in other products. That meant customers didn't always trust what they were buying; it might be gilded. The many-different-names thing—sucrose, glucose, dextrose, grape sugar, corn sugar—was an additional note of caution to consumers. It wasn't semantics. The cloudiness in terms opened the door for swindlers, accelerating the view that scientific terms were cloaking hazardous foods and fomenting the possibilities of distrust in the marketplace.

Pure food advocates tried to clarify this. Ellen Richards, a pioneer among them, acknowledged in *Food Materials and their Adulterations* (1886) the confusion "caused by the loose ways in which the term 'glucose' is commonly used." She told her readers that "formerly it was the designation of all the manufactured products, whether solid or viscous, but of late the term 'starch sugar,' or dextrose, covers the solid sugars and glucose means the syrup form, from the Greek *glukus*, meaning sweet." US Department of Agriculture (USDA) chemist and pure food agitator Harvey Wiley made the same point, admitting in his blustery way that "the term 'glucose' is used by chemists, rather too loosely [as] any sugar of the composition $C_6H_{12}O_6$." A report from the National Academy of Science in 1884 offered similarly that manufacturers and purveyors used grape sugar, glucose, beet sugar, dextrose, maltose, dextro-glucose, glycose, potato sugar, and fruit sugar interchangeably.[14]

Manufacturing scale and its resource draw were further reasons for questioning glucose. All of the examples above meant that a good deal of the worry was market-side fretting by the consumer (at the table). It also didn't help that consumers didn't know where this stuff came from (the farm). When they did learn, it wasn't too encouraging. On the production side of what was still an agrarian world, glucose manufacturing facilities seemed dangerous and wasteful. Resource demands for converting starches into sugars were immense.

The American Grocer was a trade paper representing the interests of urban grocers at the time. They were on the scene to provide an early overview of the production process. Their report came from Buffalo, where factory men raised corn from rail cars seven stories to the top of the building.

From that towering height overlooking Lake Ontario, their coworkers began a pulping process to create a mash. Through heating, filtering, and mixture with chemicals—sulfuric acid—the "pure, limpid thick syrup" sluiced through filters and boilers at stepped-down temperatures to create a final product. "This is called glucose, or grape sugar by manufacturers." In its best form it is "whiter than refined cane sugar, but less sweet."[15]

Not only were the factories outsized in structure, but I risk understatement by saying they left their marks on the landscape. They drank copious amounts of water. They hauled in tons of chemicals, literally, to convert the starch sugars into refined syrups. A new generation of chemical companies provided the sulfuric acid, but cities, municipalities, and adjacent rivers provided the water.

One Chicago factory handled 12,000 bushels of corn a day at its 1880 outset, with plans to double capacity to 25,000. A bushel weighed between 55 and 70 pounds, so on average the facility would be handling about 1.5 million pounds of corn per day. The process typically used a 3:1 ratio of water to starch. This meant that each operating day one factory consumed 4.5 million pounds, or over a half million gallons, of water. That's water for the baths of 10,000 homes from just one factory. To aid the process, chemists supplied three *tons* of sulfuric acid every day.[16]

When Nieuwland pitched the Grangers' Glucose Company in 1874 there were but a handful of factories in the United States; by the early 1880s there were twenty-nine factories in operation. In 1882, the first year the government kept such statistics on the commodity, producers shipped 5 million pounds of glucose to foreign ports. That increased more than sevenfold by 1890, to 38 million. At the turn of the century, glucose exports quadrupled again to 185 million pounds in 1900 alone. In the first five years of the twentieth century exporters maintained that 1900 average to ship 750 million pounds to foreign ports. These weren't trifling amounts. The new age of processed commodities was thriving.[17]

To achieve this level of production, glucose makers moved into the Midwest and Upper Mississippi Valley where, as Nieuwland had anticipated, a postbellum corn bounty gave farmers a market advantage over potato glucose in Europe. Yet this too agitated suspicion. After the early years on Lake Ontario it wasn't uncommon for a glucose company to incorporate in the east (often New Jersey) but to build facilities in Iowa and Illinois, far away, with little oversight on rivers that they thought nobody cared about. But people did care. The absentee relationship led to heated legal arguments over resource use back then.

In various lawsuits against a glucose industry funded by coastal elites, plaintiffs—the heartland farmers—worried that East Coast investors were poaching their supplies. In one case of water overuse, a coal and iron foundry sued the Pope Glucose Company over riparian (river) rights. Manufacturers had built a dam on Illinois's Fox River to provide mill power, but Pope's glucose factory drew so much water from the river that the mill couldn't operate. Pope was ordered to reduce its use "on nights and Sundays or at hours and times that were unusual and unreasonable."[18] In another case, this one over water pollution, the State of Iowa brought suit against the Firmenich Manufacturing Company of Marshalltown, Iowa. The state won the case for damages as the factory's waste, including "glucose, acids, sulphuric acid, sulphur and other poisonous substances," was damaging enough to make the local creek "so impure that many of the fish in the stream died."[19]

If you were one of those Grangers in the Midwest, these kinds of things put the promise of glucose in doubt. If you were a consumer on the East Coast, you might not be bothered when factories slurped up the rivers of Iowa and the Great Lakes from Buffalo to Chicago through an industrial straw. That was way over there. But what about those chemicals? Three tons of sulfuric acid a day, and what about residual acid in the eventual table syrup? That was right here on your table; that didn't sound good. The health question thus raised concerns too. Despite efforts to neutralize the acid—more factory steps and more chemistry—other chemists almost always found residual sulfuric acid in the tested end product.[20] And if residual acids and sulfites weren't offensive enough on their own, problems of digestion could be. Although *Scientific American* considered it "perfectly harmless," the editor of the *Boston Journal of Chemistry* reported digestive problems. Glucose "gives rise to flatulency, colic and Nausea," he wrote. Elsewhere, consumers were reporting that glucose gave them "flatulency and painful affections [*sic*] of the bowels."[21]

The satirical weekly *Puck* understood that dynamic. They saw glucose as a corrupt part of the candymakers' trade (figure 6.1). The doctor and

6.1 "Our Mutual Friend." This cover image from an 1885 issue of *Puck* did three things at once. It highlighted the perception that the medical and mortuary professions were the end result of eating chemical sugars. It showed the ways people thought of glucose as one among many of those chemicals. And in its title it offered a callback to the 1865 Dickens novel of the same name, *Our Mutual Friend*, which centered on the health and purity of English life amid the pursuit of money at all costs. ("Our Mutual Friend," *Puck* 16 [January 7, 1885], 1.)

the gravedigger joined hands in a partnership; glucose-based fake candy would be the death of them all.

By the 1890s, the laundry list of reasons why consumers were suspicious of glucose was as long as Nieuwland's sideburns were bushy: secrecy, fears of charlatans, naming conventions, scale and complexity, massive resource use, absentee landlord relationships, and so on. The legacy of suspicion was set in place. At the turn of the century the new product, a new ingredient, was everywhere; it was capital intensive, and it was controversial.

While consumers worried, producers found it to be handsomely profitable. The capitalists got together to ensure the profits would keep flowing like pourable syrup. Investors formed the Glucose Sugar Refining Company in 1897 as a way to aggregate their risks and advertise their might. With this, they emulated the model of the Sugar Trust. The press referred to them as the Glucose Trust almost instantly, punctuating its doubtful demeanor to readers. A group called the Corn Products Company soon took over the Glucose Trust as a way to bury the glucose name and overcome its stigma. Then in turn the Corn Products Refining Company (CPRC) bought them out. As if the entanglements of Gilded Age industry weren't problematic enough, the CPRC was headed by former executives at Standard Oil.

In 1903 the CPRC stewarded a new syrup to the market by attempting to escape the stain of adulteration through a multipronged marketing campaign. They placed ads in domestic science papers and the women's press while selling cookbooks showing how to use the syrup in recipes. They also worked hard to say "corn syrup" instead of the tainted "glucose." The syrup designation was a way to avoid the glucose label. This was like Philip Morris changing their name to Altria so if you thought Philip Morris → cigarette → cancer, then at least when cancer came around you wouldn't think of Philip Morris anymore.

The CPRC's new product was the most successful corn-based sweetener to follow the first generation of glucose. They soon gained regulatory support. I've been referring to corn syrup throughout this chapter a bit flippantly; it wasn't until 1908 that the USDA ruled, with backing from President Roosevelt, that glucose as syrup derived from corn could be labeled corn syrup.[22] This marked the end of glucose as a publicly reviled term. Legally, the fight against it as an adulterant then died a quiet death, subsumed under a proliferating industry of processed sweeteners and lost beneath a new label. The new world of processed foods was too far on its way. Thereafter the term itself fell out of favor while corn syrup, starch sugar, dextrose, and fructose became more common references. The glucose in that diabetes test your doctor orders was free to take on its modern biological reference as blood sugar.

The CPRC's new product was called Karo Syrup. They advertised it as "pure syrup" at the time, not pure sugar. This meant they took to heart the murky origins and sullied reputation of glucose by avoiding the sugar claim, but they kept the purity one. A century later the Corn Refiners Association (CRA) thought enough time had passed to sell its then-fructose-based Karo Syrup as corn "sugar."[23]

There is a history to be told about the transitions across the century that I can only nod to here, particularly as chemists had developed the high-fructose version of corn syrup (HFCS) in the 1930s.[24] This led to a false advertising ruling a decade later, pitting the makers of fructose against the makers of sucrose. The fight raged into the late century, by which point Coca-Cola had famously switched to HFCS from cane sugar. That was 1984. The US Food and Drug Administration (FDA) maintained, as its USDA forebears had three-quarters of a century earlier, that pure HFCS was syrup, not sugar. You'd be dismissive to think these semantic distinctions were trivial. On the label they were a cue that what you were consuming was other than cane sugar but still a *kind* of sugar. In the courts those distinctions were monumental; they stood for different industrial powers, different corporate alignments, different economic forces, different environmental histories.

In the early twenty-first century the CRA sought to call their fructose "corn sugar" in an attempt to overturn FDA policy. Starting in 2011, the lobby for sugar growers sued the CRA for false advertising. The CRA had launched an ad campaign for HFCS claiming that all sugars were alike, or in their words, "sugar is sugar." Their online and television ads tried to blur the distinction between sweeteners to ask consumers to trust in corn. They wanted their product, Karo Syrup, to bear the label "corn sugar," not corn "syrup."

Cane and beet sugar (sucrose) growers didn't like that. Theirs was the "real" sugar; corn syrup was not. Karo was being deceptive and dishonest. It was artificial. *Saturday Night Live*'s parody was based on the idea that a field-grown product was obviously better than a chemically processed one. That's how the court arguments went, too. The district court case in California pitted sugar against corn refiners, with plaintiffs—the sugar side—arguing against corn syrup that "HFCS is not natural but is, instead, a fabricated product requiring advanced technology." It was "the result of extensive scientific research and development." In language you might say was ripped from century-old headlines, trying to sell fructose as "real" sugar was "sneaky and dishonest."

The perception of a long con that began in the 1870s has carried through to our day. Though I don't know how you feel about corn syrup, high fructose or otherwise, I will say this: if you have reservations, if you don't trust it, if you're suspicious, you've got history on your side. To have the future on our side, escaping the folly of searching for the one pure thing is a good idea, as is attending to modes of trust. Put another way, puzzling through the cultural question of why Budweiser's ad agency could assume suspicion of corn syrup as a given will help more than strictly technical solutions to recover or reinvent a pure sugar—now Stevia, now Xylitol. Suspicious food will long remain a cultural as well as technical problem. We'd do well to shape our food policy directives with that as the core prerogative.

NOTES

This chapter is adapted with permission from *Pure Adulteration: Cheating on Nature in the Age of Manufactured Food*, by Benjamin R. Cohen, published by the University of Chicago Press © 2019. All rights reserved.

1. David Leonhardt, "Big Sugar versus Your Body," *New York Times*, March 11, 2018, https://www.nytimes.com/2018/03/11/opinion/sugar-industry-health.html. The uproar over soda taxes both proposed and enacted is but one recent issue raising those points. Corn syrup may be so bad as to be subject to a vice tax, like alcohol, cigarettes, and gambling.

2. Susan Salisbury, "Sugar Growers' Lawsuit over 'Sweet Surprise' Advertising Goes to Trial," *Palm Beach Post*, November 2, 2015, https://www.palmbeachpost.com/article/20151102/BUSINESS/812019552. See also *Western Sugar Coop., et al., Plaintiffs, v. Archer-Daniels-Midland Co., et al., Defendants*, No. CV 11-3473 CBM, February 13, 2015.

3. If you don't know Vanderbilt the railroad tycoon—they named a Nashville college after him in thanks for his money—know that he had famously big bushy sideburns. As did Nieuwland.

4. Edward Nieuwland, *The Grangers' Glucose Co.: A Very Safe and Most Profitable Investment for Farmers* (New York, 1874), 3.

5. Julius Frankel, *A Practical Treatise on the Manufacture of Starch, Glucose, Starch-sugar, and Dextrine* (Philadelphia: H. C. Baird, 1881), 196.

6. See Deborah Warner, "How Sweet It Is: Sugar, Science, and the State," *Annals of Science* 64 (April 2007): 147–170.

7. It was such a big deal we have two chapters in this volume about it—see also Singerman, chapter 11.

8. David Singerman, "Inventing Purity in the Atlantic Sugar World" (PhD diss., Massachusetts Institute of Technology, 2014), 13–14.

9. Thomas Rogers, *The Deepest Wounds: A Labor and Environmental History of Sugar in Northeast Brazil* (Chapel Hill: University of North Carolina Press, 2010), 30–32.

10. See Wendy Woloson, *Refined Tastes: Sugar, Confectionery, and Consumers in Nineteenth-Century America* (Baltimore: Johns Hopkins University Press, 2002), 24–25; Lawrence B. Glickman, "'Buy for the Sake of the Slave': Abolitionism and the Origins of American Consumer Activism," *American Quarterly* 56, no. 4 (December 2004): 889–912; William Fox, *An Address to the People of Great Britain, on the Propriety of Abstaining from West India Sugar and Rum* (Sunderland, UK: T. Reed, 1791), 4; and Kristin Hoganson, *Consumers' Imperium: The Global Production of American Domesticity, 1865–1920* (Chapel Hill: University of North Carolina Press, 2007), 121, for commentary on a 1909 effort by the National Consumers' League to campaign against slave-grown cocoa from the Portuguese islands of San Thome and Principe. See also Ethan D. Schoolman, "Building Community, Benefiting Neighbors: 'Buying Local' by People Who Do Not Fit the Mold for 'Ethical Consumers,'" *Journal of Consumer Culture* (August 4, 2017), https://doi.org/10.1177/1469540517717776.

11. Deborah Warner, *Sweet Stuff: An American History of Sweeteners from Sugar to Sucralose* (Washington, DC: Smithsonian Institution Scholarly Press, 2011), 89 and 94.

12. Glucose's common synonym, "grape sugar," followed from chemists who first derived glucose from grapes in the later 1700s. In 1811 the German-born chemist Gottlieb Kirchhoff prepared it by applying a dilution of sulfuric acid to various starches. Kirchhoff's recipe paved the way for the first half century of grape sugar production in Europe. The name was common, though rarely thereafter was the sugar derived from actual grapes. Corn, sorghum, and beets provided the raw material for most glucose in the United States, with corn in the lead; Europeans derived most of their glucose from potatoes. Other smaller producers utilized sources from the list of grain, rice, fruits and the like. Frankel, *A Practical Treatise on the Manufacture of Starch, Glucose, Starch-sugar, and Dextrine*, 196.

13. Warner, *Sweet Stuff*, 112 and 241. Warner's work provides the best summary of glucose operations at the time. This paragraph and the next are indebted to her work.

14. Ellen Richards, *Food Materials and Their Adulterations* (Boston: Estes and Lauriat, 1886), 92; Harvey Wiley, Box 1, RG 97, NARA, 8/3/1881.

15. *American Grocer*, as quoted in Wells, *The Sugar Industry of the United States, and the Tariff: Report on the Assessment and Collection of Duties on Imported Sugars* (New York: Evening Post Press, 1878), 17–19.

16. "Chicago Sugar Refinery," *The Daily Inter Ocean*, December 25, 1880, 3; "Grape Sugar from Corn," *New York Sun*, July 21, 1880. On water-to-starch ratios, 3:1 is the average of the five processes summarized by Frankel, while some used 4:1 and some 2:1. For reference, the Buffalo Grape Sugar Company handled 5,000 bushels of corn a day in 1880. For water reference, see https://water.usgs.gov/edu/mgd.html.

17. B. R. Cohen, *Pure Adulteration: Cheating on Nature in the Age of Manufactured Food* (Chicago: University of Chicago Press, 2019), chap. 6.

18. *W. H. Howell Company v. The Charles Pope Glucose Company; Charles Pope, J. Wm. Pope, Richard Pope, Arthur Pope and Edward Pope*, Court of Appeals of Illinois, Second District 61 Ill. App. 593.

19. *The State of Iowa, Appellee, v. W. S. Smith, Appellant*, 82 Iowa 423; 48 N.W. 727; 1891 Iowa Sup. May 1891. See also Warner, *Sweet Stuff*, 113.

20. Arthur Hassall, the leading purity chemist in Britain, reported that finding sulfite of lime was an easy way to tell if the sugar had been adulterated. Arthur Hassall, *Food: Its Adulterations, and the Methods for Their Detection* (London: Longmans Green, 1876), 220, 274.

21. Nichols quoted in *New England Journal of Agriculture*, April 1, 1882; the rural press is "The Glucose Fate," *Colman's Rural World*, May 26, 1881. My retrospective view tells me that the trace amounts of acid probably weren't a big deal, given that we have that acid in our stomachs anyway. It's an easy comment to make in hindsight. At the time, saying it likely wasn't a big deal would not have been a convincing counterpoint without the mantle of cultural authority to go with the claim.

22. Warner, *Sweet Stuff*, 116–121. See also *Interstate Commerce Commission Reports* 28 (1914): 675. A century later the Corn Refiners Association tried to change the name again to "corn *sugar*," but in the case of *Western Sugar Coop. v. Archer-Daniels-Midland*, lost in a confidential February 2015 settlement.

23. Salisbury, "Sugar Growers' Lawsuit over 'Sweet Surprise' Advertising Goes to Trial." See also *Western Sugar Coop., et al., Plaintiffs, v. Archer-Daniels-Midland Co., et al., Defendants*.

24. Warner, *Sweet Stuff*, helps here.

7

Modern Food as Mass-Consumed Food

THE SEARCH FOR THE AVERAGE CONSUMER
BREAKFAST CEREAL AND THE INDUSTRIALIZATION OF THE AMERICAN FOOD SUPPLY

Michael S. Kideckel

Shredded Wheat was developed in 1893 and sold as Nature's cure for a nation in dietary distress. In the period between about 1880 and 1920, many people in the United States associated food with pain. Especially those with money—but also plenty without—complained of feeling tired, hurt, bored, and constipated. Some enterprising individuals spotted an opportunity in this suffering. They offered a cure in which they linked living a "natural" life with eating "natural" food. That capital N in the first sentence, which the food's producers (along with many others) used to indicate the agency of the natural world, suggests how closely the makers of Shredded Wheat, related breakfast cereals, and the packaged foods they inspired linked their products to being "natural"—a term so disputed then and now that I put it in quotes when I use it here. For those manufacturers and advertisers, "natural" meant good, painless, not constipated. It meant items believed to be more suitable to long human life than cured meats and white bread.[1]

Among these proponents of "natural" food was Henry D. Perky. An attorney turned engineer, Perky in 1893 became one of the first people to invent and mass-market a food as being related to nature. He argued that wheat was perfect because it offered nutrients in the exact proportions that the human body needed them, and that his factory-made version of wheat, which he dubbed Shredded Wheat, was "natural" because he did not alter the grain much, did not add anything to it, and did not allow people's hands to touch it. At the beginning of the twentieth century, the hands in question—the labor in most food factories—belonged to immigrant workers. Perky linked his factory to naturalness in part by emphasizing the exclusion of people whose touch he considered a pollutant.[2]

To invent a new food in this time period, especially one that was meant to be a cure-all, meant gaining trust among people who felt increasingly disconnected. Americans lived farther from farms and extended families, out of reach of the parents, preachers, and farmers who had once guided their food decisions. Amid a series of scandals involving foods ranging from meat and milk to flour, and without significant government regulation of food safety until after World War I, it could be difficult for most customers to know whom to trust.[3] In the absence of older sources of authority on food, as Anna Zeide writes in this collection (chapter 10), trust was often supplied by celebrities, domestic scientists, and the promoters of food brands. In the same way that parents wary of unknown neighbors tell their children to avoid homemade Halloween treats, major companies in the early twentieth century assured a wary public that their name on a package guaranteed safety.

Perky was not alone in his packaged and mass-produced efforts, but his work placed him as a pioneer. He sought public trust by claiming that his goods were not only safe, but "natural." He named his company the Natural Food Company and argued that his corporate wheat biscuit, unaltered and clean due to its journey through a high-tech, hands-off factory, could cure the nation's fatigue, indigestion, and other pains.

A series of other brands followed Shredded Wheat, including notables such as Postum Grape-Nuts (1897) and eventually Kellogg's Corn Flakes (1906). By 1900 the new trend had resulted in what many observers called a cereal "boom" or "gold rush." Breakfast cereal became one of the most advertised and eaten of the factory-made, mass-distributed foods changing American diets at the turn of the twentieth century.[4] A quest for "natural" food helped industrialize the food supply by leading to some of the earliest and most consumed mass-produced packaged food.

In partnership with the mass-produced food system, then, was the idea of mass-consumed food. That was food purchased by the so-called average consumer, the elusive subject I wrangle with in this chapter.[5] Historians of food and consumption have long struggled to define and understand who consumers are and why they act. We have access to reams of financial data from companies, letters between executives, newspaper articles, and advertisements. We look for evidence of consumers' instincts, intentions, and opinions by looking at what they buy, when, how, and how much. Yet, for all the documented consumer behavior and the artifacts produced by companies, many in our field have concluded that all historians can extract from the sources is, as the historian Pamela Walker Laird put it, "what their creators thought audiences wanted to see."[6]

Customers do not, in general, write much about why they buy what they buy. Especially for the period that this book considers, before focus groups or market research surveys or even large-scale consumer activism, it is difficult for researchers to know who bought what, what those customers were like, or why they made their purchases. We have been stuck, for the most part, telling the story of advertisers, manufacturers, and prominent retailers—stories typically about the powerful white men who took credit for shaping the food supply. This limitation, however, is changing.

Thanks to newly digitized documents, historians can conduct social history more comprehensively than before. Scholars can easily find names in archival documents and then again in local newspapers, census records, and city directories. Reconstructing the ordinary consumer is an alluring prospect for historians of food eager to write more inclusive histories that offer a fuller picture of how and why the food supply has changed. Even if they don't use the term "average consumer," historians often seek anecdotes about people presumed to be representative. But with major logistical hurdles cleared to tracking the lives of humbler people, an analytical one remains: who, exactly, is an average consumer? And assuming that one can be found or defined, what does looking at their lives add to our understanding of changes in food history?

In what follows I test out some of the possibilities of historical consumer research. I use as a case study a purchaser of Shredded Wheat—a seemingly powerless person who patronized the breakfast cereal industry, which ranked among the most vocal proponents of an automated, packaged food supply at the turn of the twentieth century. Looking at the history of breakfast cereal through this "everyday person" shows that while research into typical consumers might be possible, historians have to tread carefully because the idea of an everyday person can reinforce the very hierarchies that the attempt to fill in historical absences tries to dismantle. An important component of power, after all, is the ability to define who is "ordinary" or what is "average."

Seeking the average consumer shows the opportunities and limitations of using individual behavior to understand collective trends. We know, for instance, that cereal producers spread widely the message that people should eat packaged breakfast food because it was "natural." We know, too, that many people bought cereal and its makers became wealthy. Something in the advertising seems to have worked. But the fact that many people bought cereal does not tell us whether they believed the messages that advertisers spread. And most of the stories that have gained traction around breakfast cereal offer little help in this regard, relying as they do on the trope of an eccentric genius inventor.

A common depiction, familiar through news features, trivia nights, novels, movies, and even the television series *Drunk History*, traces the origins of breakfast cereal to the campaigns for right living of John Harvey Kellogg, the doctor whose mythos Adam Shprintzen helpfully punctures in chapter 13. This origin story is not wholly false, but its emphasis and sequence make it far less accurate than the story of one of the world's most widely eaten foods should be. The narrative focuses on one character, Kellogg, and on one place, Battle Creek, Michigan, both of which played only a part in popularizing cereal. The story often identifies a motive for cereal—sexual abstinence—that had almost nothing to do with cereal's development as a mass-market good. It is incomplete in part because of its focus, which takes cereal out of its social context and frames it mostly as an oddity. The narrative also brings us nowhere near an understanding of why people bought dried grains in cardboard boxes.

Instead of the genius inventor, the story I tell here is of the otherwise prosaic Hattie Teahan. Her purchase of breakfast cereal offers an opportunity to understand how a seemingly ordinary person helped industrialize the food supply by buying one of the most heavily advertised and most widely distributed foods of the early twentieth century—one that its founders, despite proclaiming their factory production methods, branded as "natural." Teahan's life offers clues about whether she purchased cereal for the reasons that advertisers wanted her to.

The story of Hattie Teahan can help us illuminate how cereal, and by extension industrial food, became so popular. In the summer of 1908 Teahan bought more Shredded Wheat than anyone else in her small town of Charlemont, Massachusetts. She bought more by an impressive amount, taking home three times the Shredded Wheat of any other family that summer.[7] Yet Teahan makes a strong case study precisely because she resembles many people on many metrics. Like a great number of American consumers in 1908, she was white, female, lived in a small agricultural town with her husband and children, attended a Christian church, and lacked access to large amounts of wealth. The apparent averageness of Hattie Teahan is illustrative. It is also deceptive.

THE STORY OF CEREAL FOR HATTIE TEAHAN

I begin with the value of Teahan's story, which shows how individual purchases helped industrialize food, gives clues to the success of cereal advertising, and, above all, demonstrates how the industrialization of food looked different across scales of space and time. Teahan was born Harriet Hicks in 1865 in Hawley, a majority-white town in Franklin County, Massachusetts. A biographical review of Franklin County's "leading citizens" lists her father, Rufus Hicks, as "a prosperous farmer and mason."[8] At the age of twenty-three, Hattie Hicks married a thirty-year-old railroad foreman, William Teahan, of nearby Charlemont by way of Rouses Point, a town in New York State, just a mile south of the Canadian border.

Together, the Teahans performed middle-class gender roles common at the turn of the twentieth century as they raised their two children, Emma and Roscoe. Hattie Teahan kept house and maintained an active calendar, singing in her church choir and attending gatherings covered by the social pages in local papers. William Teahan worked outside of the home, on the railroad, from which he earned a comfortable living but not enough to eliminate the mortgage on the family's property.[9]

Charlemont itself was a town of just over 1,000 people, dotted with fruit trees and without a railroad until 1876, fifty years after passengers began traveling by rail elsewhere in the northeastern United States. The primarily agricultural town featured churches, a robust lumber industry, and a history of unprofitable prospecting efforts. Many found work at the nearby Davis mine, removing pyrite, or "fool's gold," from the ground. Given the small footprint of the town, it seems safe to reason that the Teahans lived close to both general stores: A. L. Avery and Son, the larger one at four stories tall and run by one of Charlemont's most prominent families, and Wells's, which a local historian remembered as the store that had introduced townspeople to bananas.[10] Charlemont is one of several towns that are known as the Hilltowns of Massachusetts. The best-known neighboring towns are Deerfield, twenty miles southeast, and Williamstown, twenty-five miles northwest. Some members of the Charlemont community were affiliated with the latter, traveling there to work at Williams College.

At some point before 1908, Hattie Teahan started buying Shredded Wheat at A. L. Avery and Son.[11] Her husband's name was left in the ledger, though historians have shown that the mere fact of his name being written down did not mean William Teahan bought anything himself. Storekeepers usually kept accounts in women's names only when those women were widowed or single.[12] Women, however, did most of the shopping. Advertisers and editors—of whom many were women—certainly believed this, and aimed most of their messages at the presumed female housekeeper.[13] There is no reason to believe that Teahan would have been an exception to this rule, and so it was almost certainly she who bought Shredded Wheat in Charlemont by the summer of 1908.

Shredded Wheat was not new in 1908, but it was not common in Charlemont. Understanding the unusualness of cereal in Charlemont as opposed to nationally shows the importance of scale to historical narratives. On a national level, Shredded Wheat and other breakfast cereals had become prevalent by 1908. Although developed only in the 1890s, as early as 1904 one seller had insisted that cereals had become "as much a necessity as bread," and other contemporaries agreed that the foods were ubiquitous.[14] Many writers linked breakfast itself to store-bought cereals, erasing the meal's prior connotations as either leftovers or a protein-rich extravagance for workers bracing for a day of physical toil. Cereal's early marketers had, in fact, tried to sell the food as more than mere breakfast. Urging housekeepers to make dishes such as "mushrooms in Shredded Wheat Biscuit baskets" and "Shredded Wheat fish chops," the Natural Food Company insisted that one could use Shredded Wheat to serve breakfast, but also as "a luncheon" or "dinner with Shredded Wheat as a basis for the various courses."[15] Even so, Shredded Wheat and other cereals came to be widely called "breakfast food," eventually adopting a role similar to the one they fill today for many people.

So Shredded Wheat was hardly new in 1908, but it still had not caught on widely in Charlemont. By that time most customers left A. L. Avery and Son with unbranded staple foods.[16] Charlemont residents' resistance to Shredded Wheat and similar products is understandable. In 1908 the food would have seemed strange: a small woven pillow from a cardboard box meant to be splashed with cream. As Zeide notes, consumers treated canned food warily at first, too, but at least canned corn and tomatoes looked like corn and tomatoes once you opened the container. Shredded Wheat, no matter its creators' insistence that it was a "natural" food, existed only inside a cardboard box. Teahan could have seen nothing more than a picture or perhaps a sample of the stuff before she decided whether or not to eat it.

Something, at some point, compelled Teahan to buy this strange food. In fact, although A. L. Avery and Son did not sell Shredded Wheat often, it did sell it in large amounts: almost all of it to just four families, and most of all to Hattie Teahan. The causes of Teahan's purchase were likely multiple. She might have seen Shredded Wheat at one of the expositions that the company held around the world, at which the Natural Food Company installed displays staffed by women in white coats giving out samples and offering cooking demonstrations. At the same time Shredded Wheat was a highly advertised food at a time when few other foods were: Teahan could have seen the cereal in ads in newspapers, magazines, billboards, or streetcars. She might have even gone to the Shredded Wheat factory in Niagara Falls, New York: 100,000 people visited the popular tourist attraction each year.[17] Moreover, Shredded Wheat was practically local food in Massachusetts, with roots in Worcester, and nearly 20 percent of Shredded Wheat in 1903 was sold by sales agents in New England—more cases per head sold than by any other agents in the world.[18] The Natural Food Company worked hard to make a strange food seem familiar, especially in New England, and at some point Teahan seems to have shown up at the general store willing to try it.

Once Teahan had entered the store, a grocer may have helped push her toward buying Shredded Wheat. Unlike today's predominantly self-service stores, in which consumers interpret labels to determine what to buy, a grocer probably staffed A. L. Avery and Son and assisted the store's customers. That grocer would have helped Teahan navigate the boxes bearing company sales pitches in varied designs, arrayed in stacks and rows along aisles and windows. Some boxes caught the eye more than others. Contemporaries described cereal boxes as garish, but one also recalled the Shredded Wheat box as a "manila brown package with only the black ink on it, [having] no display value at all."[19] Either way, the helping hand of a salesperson, whom the Natural Food Company may have personally solicited to sell its product, may have guided Teahan to Shredded Wheat.[20]

With her new purchase, Teahan would have felt relief at the prospect of not having to cook it. The coal stoves widely used at the time were hot, dirty, and prone to explosion.[21] Teahan most likely prepared food for her family herself. She must have felt resignation, resentment, and some worry about using the coal stove to cook each morning. She would have sweat through her clothes in front of the stove in July and likely felt exhausted from the morning's cook even before starting her remaining tasks and activities for the day. If she did not use the stove, even if she plucked meats from her icebox and bread from the pantry and assembled them on the table, work still marked the morning. A box of Shredded Wheat would not obviate this dreary routine, but it could help.

As they ate it, the whole Teahan family may have enjoyed the benefits of a food sold as having cooling effects. When they sat down to a bowl of Shredded Wheat for breakfast, the family would first remove the biscuits from their container. Unlike fresh bread and meats, these biscuits emitted no heat and little smell. Before eating, the Teahans probably covered the biscuits in cream, perhaps adding berries or other summer fruits. In a July heat that would rise to over 90° Fahrenheit, the biscuit doused in fresh cream no doubt felt refreshing.[22] Like eating mint or drinking ice water, the meal probably felt truly pure.

The Teahans left Charlemont in 1908. They sold their property for a small amount and moved to Mechanicville, New York, a railroad hub in Saratoga County.[23] The family was active in their new community: they joined choirs, wrote for school newspapers, and participated in youth organizations. In the clearest connection between their lives and their proclivity for Shredded Wheat, the Teahans' son, Roscoe, became a practitioner of homeopathy, one of what were then called nature cures—a form of medicine popular among those who promoted the vegetarian diets linked to breakfast cereal and "natural" food. This is the strongest clue that Hattie Teahan's story offers about an ideological link to the marketing campaigns behind Shredded Wheat. The other information about the family's life leaves an open question about whether she bought the cereal because she saw it as the "natural" food savior that Perky imagined it to be. It is possible, of course, that she just found it convenient, or that someone in the household liked the taste.

One striking conclusion from looking at Shredded Wheat in Charlemont is that something could be a major national trend while selling in relatively small quantities. A. L. Avery and Son sold more Shredded Wheat than most other branded foods, people wrote about the cereal on national scales, its advertisements appeared widely, and its makers made millions. And still, most people came out of the largest provisions store in Charlemont without touching it. The industrialization of the food supply—the making of modern food—did not have to occur evenly or to everyone, or for the same reasons. It just had to affect the lives of enough people, and enough of the sort of people who show up in store ledgers.

In this sense, the story of Hattie Teahan explains the strange process of modernizing the food supply. The adoption of breakfast cereal helped popularize an American diet rich in packaged, mass-distributed foods because millions of small consumers like and unlike Teahan made decisions about what they would buy on any given day. The "industrial food supply" suggests massive scale and unstoppable force, which seems like the opposite of something that could be created through small decisions; it suggests a machine too vast and self-powering for individual choice to matter. Yet, it was the consumers typically rendered invisible by the historical record, individuals important only in their own towns, if there, who helped fund and so enable the creation of an industrial food supply. If consumption happened at a mass scale, it was not that everyone ate from the same cereal bowl but that millions of people decided for their own reasons to eat the same product. A story about the industrialization of a country, then, is always a narrative about the buying habits of a single family. It is crucial to tell the story of ordinary people.

IN SEARCH OF THE AVERAGE

Still, too often the search for an "ordinary consumer" means looking for a representative consumer. The idea of an ordinary person that can stand in for other ordinary people presents significant problems. As a woman of middling means, Hattie was clearly less powerful than many whose voices typically appear in stories about the modern food supply. But to treat her as average, even if many shared her circumstances, reproduces the poisonous idea that to be a typical American was to be white, in a heterosexual marriage, able-bodied, to attend Christian worship, and to have children.

One problem lies in the idea of the consumer itself. The "consumer" as a person with a certain political position comes from the work of social movements, not scholars. It is a rhetorical flourish, an identity that emerged in the twentieth century to help articulate a place for more people in American capitalism. As historians such as Lizabeth Cohen and Lawrence Glickman have shown, political movements actively worked to enshrine the idea that being a "consumer" meant participation in American society in the same way that owning land once had.[24] Consumption is a near-universal experience, but "consumer" becoming a shared identity was no more inevitable than mass identification as people who pay taxes, commute, or put trash cans near the curb. In any case, consumption is so universal that it does nothing to limit the people whom "consumer" describes. Advertisers, industrialists, and investors are, just as much as Hattie Teahan, consumers. To look for the average consumer, then, is to search for a category created by political movements rather than a person who matches any particular description.

As many historians have argued, moreover, the idea of ordinariness is itself fraught with toxic moral connotations about what it is to be normal. Because the word "normal," like "natural," often conveys not only what is common but what is often presumed to be better—see the need for quotes?—it is never a neutral description. The case of Hattie Teahan makes this plain. Characterizing her as typical seems to make sense. To do so, however, makes her condition seem inevitable. It hides the context of her apparent ordinariness, which was engineered by European and then American governments through colonization policies and the mass murder, enslavement, segregation, disenfranchisement, and marginalization of most people in the Americas unlike her.

So, for instance, the population of the United States in 1900 numbered around 76 million, and around 67 million of those inhabitants were white. But this was true because Europeans killed more than that number of Indigenous people throughout the Americas before 1900; because US immigration policies prevented nonwhite people from naturalizing and, at times, arriving; because the US government gave free land to white people while hampering its sale to others; because the movement, employment, and education of white people was facilitated even in the northern United States in a way that was denied to Black and Indigenous peoples. From this vantage point, if Hattie was an average American, it was only because she was part of a privileged global minority whose relative prosperity was sponsored by the wealthiest governments in the world.

In part, the problem of Hattie Teahan as ordinary is a typical problem of who is present or absent in the historical record. Picking a consumer of Shredded Wheat out of the ledger at A. L. Avery and Son does nothing to tell the story of those who chose to stay away from breakfast food or of those whose purchase history does not survive. It does not lead to the story of Edward F. Smith, head of one of two Black families recorded in Charlemont in the 1900 census. One of Smith's daughters, Della, served in town as a live-in domestic worker for the Adams family, headed by a Black husband and a white wife. Della Smith almost certainly prepared the food for this family, but neither she nor the Adams family are recorded in the A. L. Avery and Son ledger as buying packaged food in the periods I examined.

Even if they had been in the ledger, the Smith and Adams families' lives are not recorded by contemporaries as fastidiously as that of the Teahans. In local papers that reported on church gatherings, youth club leadership, and even school plays, these longtime Charlemont residents are mostly absent. The 1900 census lists Della Smith's employer, Lawrence H. Adams, as a musician, but no record survives of his musicianship, even as many reports exist of Hattie Teahan and her children singing in church and school performances. Edward F. Smith shows up in the news only in tragic and demeaning circumstances: in 1901, his son, Edward F. Smith Jr., was arrested for allegedly killing Henry N. Warner, the wealthy white farmer for whom he worked and with whom he boarded. Despite officers reportedly believing Smith's insistence that he acted in self-defense, newspapers around the region described Smith as an "imbecile," called the incident a "brutal attack," and declined to mention Warner's race while referring to Smith as "the Charlemont negro." Following testimony from "experts," a judge declared Smith criminally insane and sentenced him to confinement at the notorious state hospital at Bridgewater.[25] Even when studying supposedly ordinary people, historians must remember that any positive or neutral appearance in the historical record is a sign of privilege. Many do, of course, and they have for decades rescued the stories of marginalized people and performed the crucial work of telling those stories. When it comes to consumer history, we must tell those multiple stories while also being conscious of the purpose of doing so.[26]

The problem of Hattie Teahan as a representative consumer comes from more than her privilege; it comes from the fact that a typical, average consumer is a category more useful to marketing executives than to historians. People act for a cacophony of reasons. It seems nearly impossible to avoid being reductive when telling the story of one person as an avatar for many, and it is difficult to know what number of people could fairly be said to be representative. And yet, telling small stories remains vital. I have no answer to this tension.

CONCLUSION

Historical context demands that the category of the average consumer needs an array of qualifiers to make it useful. The limitations seem to beg, in fact, for more aggregate storytelling. But this returns us to the situation that writers have been trying to solve, in which only the obviously rich and powerful have a voice in historical work. In scholarship, in the press, in classrooms, and talking to friends and family, we must reconcile the limitations inherent in consumer history with the need to do it; we must tell large histories without forgetting the individuals who drive them. The stories, at different scales, of industrial food and of people buying mass-produced "natural" food are as urgent as ever, and so is doing them justice.

There are many parallels between the interest in Shredded Wheat in the early twentieth century and conversations about food today. The idea of ethical buying is much advertised and widely debated, and consumer motivations remain obscure.[27] Hattie Teahan and her boxes of Shredded Wheat add a historical perspective to what can be anachronistically called green consumerism. Who bought the most heavily advertised "ethical" foods in the early twentieth century, and why? As in Teahan's day, it is mostly upper- and middle-class people who can afford "natural" food. Giants of business and pioneers of mass technology still—and increasingly—purvey these ethical products. We have gone from Shredded Wheat to Amazon-owned Whole Foods, not to mention the many grain products sold by the original cereal manufacturers, now major conglomerates, such as Post and Kellogg.

If history has such similarities, how much faith should today's advocates for fairer food production place in consumer choice? If the idea of the consumer is nebulous and historically constructed, who bears responsibility for the industrialized food system's benefits and problems—is consuming the fruits of mass production nonconsensual, as the historian William Leach argued?[28] Before historians assume that they understand why the food system took its shape and what its consequences were for the daily lives of ordinary people, it will be worth investigating those who helped give it form and finding enough stories to tell about them.

NOTES

1. Michael S. Kideckel, "Fresh from the Factory: Breakfast Cereal, Natural Food, and the Marketing of Reform, 1890–1920" (PhD diss., Columbia University, 2018). See Fouser, chapter 2 in this volume, for a fuller telling of the transition to white bread in Britain.

2. Kideckel, "Fresh from the Factory."

3. On trust, nature, and the burgeoning threat of "adulteration," see Benjamin R. Cohen, *Pure Adulteration: Cheating on Nature in the Age of Manufactured Food* (Chicago: University of Chicago Press, 2019).

4. Kideckel, "Fresh from the Factory."

5. Historians have discussed the construction of the idea of "normality" by social scientists and the issues with using biography as a method of elucidating untold stories. See, for instance, Sarah Elizabeth Igo, *The Averaged American: Surveys, Citizens, and the Making of a Mass Public* (Cambridge, MA: Harvard University Press, 2007); David Nasaw, "Introduction," *American Historical Review* 114, no. 3 (June 1, 2009): 573–578, and the roundtable of which this is a part.

6. Pamela Walker Laird, *Advertising Progress: American Business and the Rise of Consumer Marketing* (Baltimore: Johns Hopkins University Press, 1998), 39.

7. A. L. Avery & Son, "Journal Y, 1908–1909" (1909), vol. 44, A. L. Avery & Son Business Records, Baker Library, Harvard Business School.

8. *Leading Citizens of Franklin County, Massachusetts* (Boston: Biographical Review, 1895), 646.

9. "1900 United States Federal Census," Ancestry Library, accessed March 21, 2018, https://www.ancestrylibrary.com/interactive/7602/4113818_00431?pid=22617038 &backurl=https://search.ancestrylibrary.com/cgi-bin/sse.dll?db%3D1900usfedcen% 26h%3D22617038%26indiv%3Dtry%26o_vc%3DRecord:OtherRecord%26rhSource %3D60525&treeid=&personid=&hintid=&usePUB=true&usePUBJs=true.

10. Allan Healey, *Charlemont, Massachusetts: Frontier Village and Hill Town* (Charlemont, MA: Town of Charlemont, 1965).

11. The registers are extensive. I have seen the Teahans buying Shredded Wheat as early as 1900, but it may well have happened earlier.

12. Joan Hollister and Sally M. Schultz, "The Elting and Hasbrouck Store Accounts: A Window into Eighteenth-Century Commerce," *Accounting History* 12, no. 4 (November 2007): 426.

13. See Ellen Gruber Garvey, *The Adman in the Parlor: Magazines and the Gendering of Consumer Culture, 1880s to 1910s* (New York: Oxford University Press, 1996); Laird, *Advertising Progress*; Daniel Delis Hill, *Advertising to the American Woman, 1900–1999* (Columbus: Ohio State University Press, 2002); Sharon M. Harris and Ellen Gruber Garvey, eds., *Blue Pencils & Hidden Hands: Women Editing Periodicals, 1830–1910* (Boston: Northeastern University Press, 2004); Juliann Sivulka, *Ad Women: How They Impact What We Need, Want, and Buy* (Amherst, NY: Prometheus Books, 2009).

14. "The Cereal Trade (May 1904)," *Flour and Feed* 4, no. 2 (May 1904): 43.

15. Henry D. Perky, *The Vital Question and Our Navy, 1898*, 3rd ed. (Worcester, MA: H. D. Perky, 1897), 11.

16. On the stores in Charlemont, see Healey, *Charlemont, Massachusetts*; A. L. Avery & Son, "Journal Y, 1908–1909," 7.

17. Allison C. Marsh, "The Ultimate Vacation: Watching Other People Work, a History of Factory Tours in America, 1880–1950" (PhD diss., Johns Hopkins University, 2008).

18. Kideckel, "Fresh from the Factory."

19. George W. Hopkins, *Post-Graduate Course: Copy* (New York: Advertising Club of New York, 1925), 11.

20. On concerns over the trustworthiness of grocers, see Cohen, *Pure Adulteration*, 45–74.

21. Katherine Leonard Turner, *How the Other Half Ate: A History of Working-Class Meals at the Turn of the Century* (Berkeley: University of California Press, 2014).

22. Teahan would have known it was 92 degrees on July 8, for example: U.S. Department of Commerce, National Oceanic & Atmospheric Administration, National Environmental Satellite, Data, and Information Service, "Record of Climatological Observations, Elev: 700 Ft. Lat: 42.6000° N Lon: −72.8333° W, Station: Fitchburg A, MA US USC00192810, 07/01/1908-08/31/1908," Report generated July 9, 2018.

23. Paul Loatman, *Mechanicville*, Images of America (Charleston, SC: Arcadia Publishing, 2013).

24. Lizabeth Cohen, *A Consumer's Republic: The Politics of Mass Consumption in Postwar America* (New York: Vintage Books, 2004); Lawrence B. Glickman, *Buying Power: A History of Consumer Activism in America* (Chicago: University of Chicago Press, 2009).

25. "Smith Held in $5000," *Boston Globe*, December 30, 1901; "Warner Dead," *Fall River Globe*, December 31, 1901; "Believe Smith's Story," *North Adams Transcript*, January 9, 1902; "Diary of Events," *St. Johnsbury Caledonian*, January 8, 1902; "Held for Manslaughter," *North Adams Transcript*, January 17, 1902; "Smith Sent to Bridgewater," *North Adams Transcript*, March 18, 1902; "Slew Employer," *Boston Globe*, July 28, 1904.

26. On assumptions of whiteness in consumer history, see also James C. Davis, *Commerce in Color: Race, Consumer Culture, and American Literature, 1893–1933* (Ann Arbor: University of Michigan Press, 2007).

27. See, for instance, Dietlind Stolle and Michele Micheletti, *Political Consumerism: Global Responsibility in Action* (New York: Cambridge University Press, 2013).

28. William Leach, *Land of Desire: Merchants, Power and the Rise of a New American Culture* (New York: Pantheon Books, 1993).

8

Modern Food as Uprising

EAT THE RICH
RADICAL FOOD JUSTICE IN MEMPHIS AND CHICAGO

Faron Levesque

Hot trash, charred pig, Froot Loops.

I take the long way to work. Head south on Stonewall, our street, and zigzag past one of the only gay bars left. I scoot up and over to Lamar Avenue, where these smells beckon and the potholes try to suck you into the sandy underbelly of this old broken river city.

The swamp heat swirling up and over the bluffs always seems to carry a wallop of a week's worth of trash coming at you from trucks full of maggots and garbage juice, moving fast and furious through cul-de-sacs and dead ends. It's a visceral reminder of the two sanitation workers, Echol Cole and Robert Walker, who were crushed to death in 1968, spurring protests for higher wages and better working conditions.

Windows down, moving low and slow, I breathe in the plume-y ghosts of smoked meats as they swirl above one of Memphis's most sacred churches: Payne's Bar-B-Q. I say a little prayer for the eternal loves and labors of Ms. Flora Payne and keep rolling.[1] The clouds of steady-burning hickory coals fade far too quickly.

The saccharine storm fronts coming out of the Kellogg's factory bring me back to the present as I drive down Airways Boulevard. I'm almost to the farm. Kellogg's in big, blood-red, swirly, glowing cursive marks the spot where factory workers produce the Froot Loops, Raisin Bran, and Rice Krispies cash cow brands on an endless loop of long shifts.

I'm from here. Unlike a historical placard or a monument cast, the stank of Memphis is unofficial and at times, like most historical nose triggers, a highly subjective marker that allows me to add some texture to the food work that I do, to get real about the way we find the past in our food.

Mirroring the national crises of memory and surge of new (old) white nationalisms, like other cities across the United States, Memphis is at a point of reckoning with the politics of memory through the removal of monuments to the Jim Crow South.[2] Activists and historians are reshaping the landscapes of North American cities through a collective mobilization around public memory.[3]

A monument to People's Grocery and the violence perpetrated there in 1892 is one such site of commemoration. It's the memory of what we now consider matters of equity in the food system that drew me to the monument.

The long history of food justice—to use the current term from organizers and scholars—is tied up with anti-lynching campaigns and anarchism; it is the struggle for Black liberation through radical economic justice and militant organizing from the field to the neighborhood; it was present at the end of the nineteenth century in ways that laid the groundwork for the next century of attention to justice in our food system.

The People's Grocery monument sits on the corner of Walker and Mississippi in front of Sam's Food Mart and just past The Four Way, an old-school soul food hub that served Dr. Martin Luther King Jr. and his comrades. I get to take some folks who are part of the community teaching kitchen I run here in Memphis around town. We call ourselves The Cornbread Academy. We wander the local foodways and food system looking at the ways that land, food, culture, and people swim together in all the muds of Memphis, grasping for some of that dirty dirty terroir.

When we come to the People's Grocery site, we read the text of the placard:

Thomas Moss, Calvin McDowell, and Will Stewart, all African-Americans and co-owners of People's Grocery (located at this site), were arrested in connection with a distrubance [sic] near their store. Rather than being brought to trial, they were lynched on March 9, 1892. Moss' dying words were, "Tell my people to go west—there is no justice for them here." This lynching prompted Ida B. Wells, editor of *Memphis Free Speech* to begin her anti-lynching campaign in this country and abroad.[4]

Another marker on Beale Street downtown adds to our memory of Ida B. Wells-Barnett. It speaks to the role both political education and food access and contested food spaces like People's Grocery played in the fight for racial justice in the Jim Crow South.

It reads, "Ida B. Wells crusaded against lynchings in Memphis and the South. In 1892 while editor of the *Memphis Free Speech and Headlight*, located in this vicinity, she wrote of the lynching of three Black businessmen. As a result, her newspaper office was destroyed and her life threatened."[5]

Wells spoke to the convergences of the struggle against lynching and for radical economic justice only a few years after her contemporary, Lucy Parsons, did similar work in Chicago. For Parsons the site was Haymarket Square and the resistance was led by anarchists, labor syndicalists, and hobohemians. The dynamic for fair labor practices—the eight-hour workday, in this case—was anchored to the exploited labor of farm industries and food provisioning.

To talk about food is to talk about power. Food has always been a vehicle for protest. It is always, already political. It was Rousseau, after all, who barked "Eat the rich" during the women-led protests over bread during the French Revolution. This story, however, is not about Rousseau.

A century later, in the 1880s, the widespread social, economic, and political upheaval brought about this time by industrial capitalism was starkly visible in the transformation of food systems. We see this in the contracted time and space in part I of this book, and we find it with the scientific aspects of those transformations in part III. We see it here with the fractured modes of trust and the interactions that new kinds of workers, activists, and eaters experienced through the upheaval.

This is a story about the politics of hunger, memory, and revolt that first took its modern shape in the late nineteenth century.

Brutal, raging ecologies of itinerant activism, revolt and resistance, relationships and solidarities: these are what connect Memphis and Chicago. People and their movements and migrations. Lucy and Ida. Formations and strategies. Cooperatives and mutual aid. Political education and direct action. All the pains and pleasures of worldmaking.

Lucy Eldine Parsons and Ida Bell Wells, Haymarket and People's Grocery: two legendary events, two legendary, radical women of color at the vanguard of what we now call the movement for food justice in the United

States. Both women fled lynch mobs in the South. Parsons came from Waco, Wells from Memphis. Central to both Wells's and Parsons's radical imaginations and activist agendas were experiences with police violence.

To understand the history of modern food is to understand the role that modern policing has played in establishing and maintaining the architecture of white supremacy, an architecture that structures oppression, some of which is centered in American food systems. Parsons and Wells spurred uprisings that sought to dismantle these supremacies of racial capitalism.[6]

Framing similar issues today through a food justice lens means recognizing the origins of food workers who took up arms and mutual aid on the shores of the Mississippi River and Lake Michigan more than a century ago.

LUCY PARSONS AND HAYMARKET SQUARE

Hunger was a flame licking a stick of dynamite in Lucy Parsons's revolutionary imagination. Dispossessed and disinherited workers could blow up the coffers of the industrialists and robber barons of Gilded Age Chicago and feast on their excesses and exploits. In a column published in 1884, the *Chicago Tribune* assured Chicagoans that there was an easy solution to panhandlers: "When a tramp asks you for bread, put arsenic on it and he will not trouble you anymore."[7]

Chicago in 1884. Two years before a riot in the Haymarket would change her life forever, Parsons wrote a manifesto in the inaugural issue of the anarchist publication *The Alarm*. She called on "each of you hungry tramps who read these lines" to "avail yourselves of those little methods of warfare which Science has placed in the hands of the poor man." Taking up arms was, for many anarchists, an increasingly necessary strategy in the face of violent industrial expansion. "Learn the use of explosives!" Lucy Parsons wrote in the final line of her love letter to the tramps.[8]

Anarchists brought their revolutionary imaginations to life in late nineteenth-century Chicago. Amid sprawling radical enclaves of immigrant activists and Black insurgents, Parsons honed a practice of freedom based on some pretty clear-cut anarchist tenets: direct action, political education, and mutual aid instead of charity.

Parsons's activist life was expansive. Many historians, however, have relegated her to "wife of . . ." status in the larger canon of American radicalism and anarchism. From her early time in the 1870s working as a dressmaker and organizer for the Chicago Working Women's Union and the Socialist Labor Party, she was a prolific writer, orator, and anti-capitalist agitator. The execution of her husband, Albert Parsons, and the seven other Haymarket "Martyrs," and the unrelenting surveillance that stemmed from the racist, state-sanctioned police violence in Chicago would haunt her days. As Angela Davis writes, however, Lucy Parsons was "far more than a faithful wife and angry widow who wanted to defend and avenge her husband."[9]

The radical subculture of Chicago was crucial to Parsons's activism. But her time moving through the south and the Texas borderlands was also formative and traumatic. Though her early life remains difficult to piece together, a handful of biographers agree that Lucia Eldine Gonzalez was born in Virginia in 1851 and was of Mexican, African, and Native American descent. During the Civil War, Lucy Parsons and her mother and brother "endured a brutal wartime 'middle passage' from Virginia to Central Texas" before she met her partner in crime and anarchy, Albert Parsons, in Waco in 1870.[10]

A former confederate soldier, Albert Parsons became a white rabble-rouser and "race traitor." As a reporter at the helm of the *Waco Spectator*, he attempted to reveal the project of whiteness at work following the Civil War by targeting the increasingly visible and empowered Ku Klux Klan in Texas. Because of the stringent miscegenation laws of the time—white supremacist laws that sought to keep white and nonwhite folks from marrying or living together—their marriage was likely extralegal. The pair's interracial marriage and antiracist politics riled the local white terrorists and soon the two were hustling their way up north to Chicagoland.

Once in Chicago, Lucy Parsons worked as a dressmaker and, through that labor, organized the Working Women's Union. She was also a founding member of *The Alarm*, a widely read anarchist newspaper. In the decade leading up to the Haymarket Affair, Parsons was honing her critical lens, writing prolifically, and standing upon many a soapbox, generating rally cries for women workers who were exploited by wealthy industrialists that anarchist Hippolyte Havel would later refer to as "the porkocracy of Chicago."[11]

Food, food production, and food provisioning were not sidelines to the anarchist movement or to the struggles of workers demanding fair wages and eight-hour days. Rather, food was a central element that grounded those movements. That grounding came about in various forms. The meatpacking facilities surrounding the area were one way to see it; the McCormick Reaper workers assembled to boost the agricultural productivity that was driving the food system were another. The food wagons that filled Haymarket Square to feed hungry laborers (and then protestors) were, too.

In the spring of 1886, between April 25 and May 1, there was a massive strike wave in Chicago. The Eight Hour Movement reached a fever pitch, with workers of all stripes aligning under one cause. The Central Labor Union rallied on the lakefront on Sunday, April 25. It was a massive demonstration where an estimated 25,000 people occupied the lakefront, toting placards that read "Private Capital Represents Stolen Labor," "Eight Hours—Working Time, May 1, 1886," and "Down with Throne, Altar, and Money Bags."[12]

Haymarket Square was a farmers' market. One of Chicago's largest open-air markets, it hosted a sprawl of wagons situated along commission houses that traded in fruits and vegetables. As a city food space, it served a few different functions. In a process of distribution much like the urban pushcarts Tom Okie describes in chapter 5, farmers would arrive in the early mornings, back their wagons full of fresh produce up to retailers who might buy veggies in bulk, then see those retailers sell to customers at stalls they rented in Haymarket Square. Others might set up ad hoc curb markets to forgo renting a stall, creating their own food hub in racially segregated space.

Striking workers used the Haymarket as a commons not only for buying food and peddling wares, but also for stirring up momentum and agitating on the side of labor.[13] On May 4, 1886—three years before the People's Grocery in Memphis, several states away, opened its doors—Lucy Parsons was thirty-five. She and a cadre of labor syndicalists held a rally in solidarity with striking McCormick Reaper Works workers in the Haymarket.

Police loomed and the protestors were on edge. As the rally ended and people were dispersing, someone—likely a Pinkerton infiltrator in the crowd—threw a bomb into the crowd. Police immediately set their targets on well-known anarchist incendiaries and journalists in the area. Albert Parsons, August Spies, Adolph Fischer, George Engel, Louis Lingg, Michael

Schwab, Samuel Fielden, and Oscar Neebe, all prominent Chicago-based anarchists, were arrested for inciting a riot in the Haymarket. The collective trauma that immigrant communities experienced following the Haymarket Affair was enduring, as was the state violence that the anarchists' executions sanctioned. This anti-radical, anti-immigrant legal architecture would prove to be part of a much larger and lasting project of white supremacy in the United States.

Writing in *Industrial Worker* on May 1, 1912, Lucy Parsons described the events leading up to the Haymarket Affair. On the afternoon of May 3, she recalled, there was a groundswell of striking workers in southwest Chicago, many of them McCormick Reaper Works employees. In the mainstream papers there was talk of riots. The police showed up in full force. "On this occasion," writes Parsons, "they shot seven working men and clubbed many hundreds unmercifully. The next evening the Haymarket meeting was called. The Haymarket meeting is referred to historically as 'The Haymarket Anarchists' Riot.' There was *no riot* at Haymarket except a police riot."[14]

The Haymarket Affair is rightly understood by historians as a signature moment in a longer history of labor strife and radical practice, yet it was also structured through the foodways of the 1880s and the work of food provisioning so crucial to urban livelihoods in ways that have often been left unstated. The Haymarket was, after all, a farmers' market, and protesters used it as the stage for their opposition to exploitative industrial agriculture as factory owners laid the foundation of what we know today as Big Ag.

Oppositions to the white nationalist legal architecture—the trial, anti-immigrant language, the state violence—were also projects of citizen-making through food. This was another way Parsons was building attention to food justice in ways that still resonate today. She saw that the state-making ambitions of radical suppression were efforts to control both labor and access to food.

The Haymarket Affair, however, was just the beginning for Lucy Parsons. Through and beyond that traumatic event she continued to refine her critique of capitalism and its many tentacles with a particular emphasis on mutual aid and revolutionary unionism.

Parsons was critical of the growing industry of philanthropy and charity as a cog in the great "wheel of fortune," especially in the face of devastating economic downturns and depressions. It was the working class that bore the brunt of the whims of Wall Street. Walking the streets of Chicago regularly, Parsons would report back in the columns of radical newspapers like *The Demonstrator*. "The free coffee wagons and soup kitchens are in full operation, and all the police stations and cheap lodging-houses are filled to suffocation." To be clear, these humble, free food efforts were not the problem for Parsons. "Charity is the dope being handed out by the robber class at present to the poor people to keep them quiet, and it is successful at least for the time being."[15] For Parsons, capitalism was the problem and charity, through an ever-expanding philanthropic industrial complex, was not a solution.[16]

Anarchists like Parsons embraced the idea of solidarity, not charity. In her work on anarchist women and diasporic radicalism, historian Jennifer Guglielmo explains that anarchist women's activism was buoyed through mutual aid networks that would enable them to "provide for their collective needs, build community, and confront their marginalized position within the capitalist world order."[17] Mutual aid is the idea that the philanthropy of the wealthy—charity—masks the exploitation of workers through benevolent capitalism and violent patriarchy. Mutual aid, in contrast, similar to the cooperative model of ownership that we see at People's Grocery, was a radical approach to collective care and community-based political engagement.

Movement strategies like direct action and labor syndicalism also rounded out Parsons's visionary politics of hunger. She was a cofounder of the Industrial Workers of the World (IWW) in 1905 and took over as editor of the anarchist newspaper *Liberator*, calling for "One Big Union!"[18]

There is so much more to say about Lucy Parsons and her vast public life of resistance and revolutionary politics. For now, however, it's worth noting that on February 23, 1941, Parsons would give her last large public speech to the Farm Equipment Workers Union of International Harvester—the new corporate face of the former McCormick Reaper Works, coming full circle and calling on workers to defend their labor and right to a full life. At the 1941 May Day Parade Parsons waved her way through the streets of Chicago atop the Farmer Equipment Workers' float one last time.

"Think clearly and act quickly," Parsons wrote, "or you are lost."[19]

IDA B. WELLS AND PEOPLE'S GROCERY

Coppery like a penny, thick like bad molasses, even a little gamey like a possum.

The white conductor's blood in her mouth probably didn't taste good, but it probably didn't taste bad either. Ida B. Wells sat firmly while the Memphis streetcar man gripped her body, tried to forcibly remove her from the "First class ladies car" on a train from the Poplar Station to Northern Shelby County. Wells took a bite out of that guy until he "bled freely," he would later plead in court. Wells sued Southwestern Rail and won a $500 settlement. The ruling, however, was ultimately overturned by the Tennessee Supreme Court.[20]

Ida B. Wells occupied that seat on May 4, 1884 exactly two years before the Haymarket Affair. It was 1884 and she had just moved up to Memphis. Born about an hour southeast in Holly Springs, Mississippi, in 1862, Wells lost her parents and young brother to the devastating 1878 Yellow Fever Epidemic. Her parents were involved with Reconstruction-era politics and the democratization of education; their daughter would carry on that mantle as a radical teacher in her own right. She studied at the freedmen's school, Rust College, at Fisk University during the summer in Nashville, and also at LeMoyne Owen College in Memphis.[21]

The loss of both her parents put Wells in the position of taking care of her five siblings. The 1878 Yellow Fever Epidemic ravaged the South and especially the port cities and distribution hubs up and down the Delta. Over 4,000 died in New Orleans, 1,000 in Vicksburg, and over 5,000 in Memphis. Rich white people had the means to flee the hot spots, but African American communities in larger cities sheltered in place and created disaster relief and mutual aid networks for the poor and vulnerable who did not have steady employment or housing and for those who lacked ventilation or sewage systems and access to safe water sources.[22]

The history of the food justice movement is the history of mutual aid. Benevolent associations and mutual aid societies, sometimes referred to as secret societies, had a long history in many antebellum southern cities. Sometimes they were aligned with churches, sometimes they were not. What united most was a commitment to shared resources, autonomous care, and collective power. This took the form of financial assistance through death benefits, insurance, health care, and vittles for the poor and hungry.[23]

Toward the end of the 1880s, strike waves rolled and disenfranchised laborers regrouped and reorganized labor and communities, as we saw in Chicago. There was an upswing in the formation of labor unions, alliances, and food and farm cooperatives during this period nationwide, with the unifying goal to foster solidarity, land justice, and economic power in new ways.[24] For example, between 1889 and 1892 the Colored Farmers' Alliance would amass over 1 million members. The cooperative union became a clearinghouse for organizations that preceded it, like the Colored Farmers' Union, the Cooperative Workers of America, and Colored Wheels. Some chapters had publishing wings; many led alternative academies in farming, offered debt relief and mortgage loan assistance, and created opportunities for cooperative exchange and distribution through the original food hubs.[25] This was the new world of reorganized food provisioning and labor that shaped Parsons's Chicago and Wells's Memphis.

When it came time to bring those farm products to market, the nineteenth-century grocer had control over transactions, with access to goods behind the counter mediated by the counter space that separated the buyer from the merchant.[26] The grocer and counter clerks would take down orders in pencil on the back of the sacks, grind coffee beans, weigh out sugar and flour, and so on.[27]

It wasn't until 1916 that Piggly Wiggly opened on 79 Jefferson Avenue in Memphis. It was the first fully self-service grocery store and precursor to the supermarkets of today. Unless you found yourself in an open-air food space like the Haymarket, a customer relied on an exchange with the grocer and clerks.

Memphis was number five on the list of the biggest wholesale grocery markets in the United States at the end of the nineteenth century.[28] The white grocer William Barrett operated what was first the only grocery in the Curve neighborhood of Memphis. Barrett was taking advantage of the access deficit in the outskirts of the city. This is a glimpse into the early history of what I will call *food apartheid*, or what some policymakers would call *food deserts*. In the words of writer and farmer Leah Penniman, the term *food apartheid* "makes clear that we have a *human-created* system of segregation that relegates certain groups to food opulence and prevents others from accessing life-giving nourishment."[29] The saga of People's Grocery shows how entrenched the history of human-made food apartheid is.[30]

In what seems to be the only existing photograph of People's Grocery, a dark horse pulls a Van Buren Cigar wagon and the driver is perched and looking directly at the camera. In the background we see a storefront with three windows and a big sign with PEOPLE'S GROCERY emblazoned. If only we could see inside those windows. It may have looked like William Barrett's store on the inside in some ways, yet by virtue of its more cooperative structure the store differed because it restructured food access and community as well as the relationship between the grocer and customer.

Many of the working Black folks in the Curve and beyond contributed capital to collectively and cooperatively sustain People's Grocery, opening it in 1889 as an alternative to William Barrett's store. Writer and sociologist Zandria Robinson writes that People's Grocery was also a beloved "destination for black people inside and outside the city."[31] Co-owned by ten Black residents in the neighborhood, it was a financially successful food cooperative held in high regard by the Black community. Tommie Moss, Ida B. Wells's friend, was the store's director. Black folks were building power by owning their labor, claiming space, and documenting the terrors of white supremacy in the ever-expanding Bluff City sprawl.

Elite and working-class white Memphians alike invested in maintaining racial segregation through the changing neighborhood landscapes of the city in the 1890s. Historian Paula Giddings describes the racial local geographies of the neighborhood as they began to shift and as white communities began laying the groundwork for Jim Crow: "Whites were establishing racially exclusive areas within the city and abandoning outlying areas like the Curve to a large number of the poorer African American migrants who joined the biracial working class already residing there."[32] These shifts in landscape also included attempts to maintain exclusive white ownership and control of spaces for food provisioning like the grocery store.

In 1889, People's Grocery was the first Black-owned cooperative grocery in Shelby County's fourteenth civil district on the then outskirts of Memphis. Within three years the primary co-op organizers and owners were killed by a white lynch mob who accused Moss of "plotting a war against whites."[33] The story of People's Grocery is a foreboding of future actions of white supremacist terrorism against Black businesses in East St. Louis, Illinois; Elaine, Arkansas; and Charleston, Houston, and Knoxville, as well as the destruction of "Black Wall Street" in the Tulsa Massacre of 1921.[34]

Anxious white business owners in Memphis were growing increasingly threatened by Black community power. Like in Chicago, Black community control over land and autonomous food provisioning was revolutionary and thus a target for state violence. White city officials and landowners codified segregation through attempts to control not only the ballot box but private and public spaces like grocery stores and curb markets. These same white boosters, property owners, and police units also relied on mob violence and the horrific public spectacles of lynching in their attempts to destroy Black power in Memphis.

Wells was thirty years old when the mob of white terrorists murdered her friends and destroyed the beloved People's Grocery. "This," Wells wrote, "is what opened my eyes to what lynching really was."[35]

While Wells was on a trip to New York, a local white-run Memphis newspaper wrote a smear piece and stoked the fires of violent white terrorists who threatened her life and burned down her newspaper office on Beale Street. Wells was writing prolifically about People's Grocery in her newspaper *The Memphis Free Speech and Headlight*. She was shining the light on the targeted trauma inflicted by white vigilantes and the police as they tried to sustain a white-knuckle grip on the south.

Joining the other 5,000 Black people using, as Zandria Robinson describes it, "their economic and political power as citizens, laborers, and consumers to vote with their feet," Wells left Memphis.[36] Like Lucy Parsons, Wells chose exile. As an itinerant activist, Wells would build and transform vast networks of political education throughout the southern diaspora and on an international scale.

Less than a month after the People's Grocery murders, Lucy Parsons wrote an editorial in the radical publication *Freedom*:

Never since the days of the Spartan Helots has history recorded such brutality as has been ever since the war and as is now being perpetrated upon the Negro in the South. . . . The whites of the South are not only sowing the wind which they will reap in the whirlwind, but the flame which they will reap in the conflagration.[37]

Ida B. Wells traveled across the country and the globe to talk about white racial terrorism and municipal governance in all of its clutchy fragility and bloodthirsty desperation. Core to her anti-lynching crusade was the evisceration of the rape myth, exploding the idea that lynching was always retribution for white Southern womanhood. It was the violence at People's Grocery that would reveal these truths to Wells.[38]

CONCLUSION

What do we know to be true of food justice after glimpsing the lives of Lucy Parsons and Ida B. Wells? Both had direct experience with police brutality, racial capitalism, and state violence. Historians have written about both women as shining stars in anarchist and anti-lynching-movement history. Parsons and Wells, the Haymarket and People's Grocery, also show us how food provisioning and access, equity and inclusion are not only definitive of the American food system, but of the history of modern food as we know it.

If we understand food systems to be the complex structures of production, processing, packaging, distribution, consumption, and recovery, then it's possible to not only map out the ways in which food has been used as a tool for maintaining social hierarchies and systems of oppression, but to also trace the shadow systems and underground economies that we see working within and without that same system. The Haymarket and People's Grocery are two sites where those alternative approaches took shape. The Haymarket was a farmers' market; People's Grocery was a co-op.

Through their activism, Wells and Parsons were stewards and storytellers who gave life—through their experiences of death—to a radical organizing tradition. The power of collective action created the conditions for disinherited folks to seize the means of food. I'm reminded of the visionary thinking of food scholar Monica White, who said, "If people can control access to food, then can we also think about community control of schools? Can we think about community policing?"[39] Parsons and Wells, like White, had the breadth of abolitionist vision to imagine otherwise.

Ida B. Wells and Lucy Parsons led the way in rethinking the axes of power around food and life. Though their experiences were different, the provisions they have left us could fill the larders of contemporary food freedom movements. They offer models of mutual aid, direct action, and political education that remind folks how to bear witness to the violence of racial capitalism while imagining what kind of strategic insurgencies it takes to get free. What endures is a usable past: a worldmaking model that challenges predatory racial capitalism through what we now call food justice.

Eat the rich.

Rousseau might have coined it, but Lucy Parsons and Ida B. Wells and many more food justice activists gave it teeth.

It's not just a worldview, theory, or manifesto.

It's an acquired taste.

NOTES

1. For more on the life of Ms. Flora Payne, see the interview conducted in 2008 for the Tennessee BBQ Oral History Project out of the Southern Foodways Alliance, https://www.southernfoodways.org/interview/paynes-bar-b-q/.

2. Shelby County Commissioner Tami Sawyer and the Lynching Sites Project of Memphis are doing local movement work in Memphis. See Laura Wamsley, "Finding a Legal Loophole, Memphis Takes Down Its Confederate Statues" *NPR*, December 21, 2017, https://www.npr.org/sections/thetwo-way/2017/12/21/572654031/finding-a-legal -loophole-memphis-takes-down-its-confederate-statues; and Maya Smith, "Lynching Sites Project Adds More Memphis Markers," *Memphis Flyer*, July 26, 2018, https://www .memphisflyer.com/memphis/lynching-sites-project-adds-more-memphis-markers /Content?oid=14781722.

3. A note of gratitude to historian and scholar-activist Doria Dee Johnson, who was at the forefront of this national reckoning. She was a beloved mentor and critical voice and force in the struggle for Black Liberation.

4. "People's Grocery," Southeast Corner of Mississippi Blvd and Walker, Tennessee Historical Commission, 4E-106.

5. "Ida B. Wells 1862–1931," Southwest Corner of Beale and Hernando, Tennessee Historical Commission, 4E-85.

6. A proper list of scholar-activists who have contributed to the literature on racial capitalism far exceeds this space. Here are a few of my favorites: Ruth Wilson Gilmore, *Golden Gulag: Prisons, Surplus, Crisis, and Opposition in Globalizing California* (Berkeley: University of California Press, 2007); Ashanté M. Reese, *Black Food Geographies: Race, Self-Reliance, and Food Access in Washington D.C.* (Chapel Hill: University of North Carolina Press, 2019); Keeanga-Yamahtta Taylor, *From #BlackLivesMatter to Black Liberation* (Chicago: Haymarket Books, 2016) and *Race for Profit: How Banks and the Real Estate*

Industry Undermined Black Homeownership (Chapel Hill: University of North Carolina Press, 2019); Cedric J. Robinson, *Black Marxism: The Making of the Black Radical Tradition*, 2nd ed. (Chapel Hill: University of North Carolina Press, 2000).

7. "More Dangerous than a Thousand Rioters": The Revolutionary Life of Lucy Parsons," *The Nation*, November 15, 2016, https://www.youtube.com/watch?v=m1AOZlUflgo.

8. Lucy E. Parsons, "A Word to Tramps," *The Alarm* 1, no. 1 (October 4, 1884): 1.

9. Angela Y. Davis, *Women, Race & Class* (New York: Vintage Books, 1981), 89.

10. Jacqueline Jones, *Goddess of Anarchy: The Life and Times of Lucy Parsons, American Radical* (New York: Basic Books, 2017), xiii.

11. Hippolyte Havel, "Kropotkin the Revolutionist," *Mother Earth* 7, no. 10 (December 1912).

12. Paul Avrich, *The Haymarket Tragedy* (Princeton, NJ: Princeton University Press, 1984), 204.

13. Food scholar Monica White discusses how the history of food spaces as commons reveals how disinherited communities mobilize justice. White, *Freedom Farmers, Agricultural Resistance and the Black Freedom Movement* (Chapel Hill: University of North Carolina Press, 2018), 8–9.

14. Lucy E. Parsons, "The Eight-Hour Strike of 1886," *Industrial Worker*, May 1, 1912.

15. Lucy E. Parsons, "Wheel of Fortune," *The Demonstrator*, January 16, 1908.

16. For a more contemporary take, see INCITE! Women of Color Against Violence, *The Revolution Will Not Be Funded: Beyond the Non-Profit Industrial Complex*, 2nd ed. (Durham, NC: Duke University Press, 2010).

17. Jennifer Guglielmo, *Living the Revolution: Italian Women's Resistance and Radicalism in New York City, 1880–1945* (Chapel Hill: University of North Carolina Press, 2010), 63.

18. Lucy E. Parsons, "Speech to the IWW in 1905," The Anarchist Library, https://theanarchistlibrary.org/library/lucy-e-parsons-speech-to-the-iww-in-1905.

19. Gale Ahrens, ed. *Lucy Parsons: Freedom, Equality & Solidarity, Writing & Speeches, 1878–1937* (Chicago: Charles H. Kerr, 2004), 21, 38.

20. Paula Giddings, "Ida B. Wells: Journalist, Civil Rights Hero, and Posthumous Pulitzer Prize Winner," *Ms. Magazine*, May 11, 2020, https://msmagazine.com/2020/05/11/ida-b-wells-journalist-civil-rights-hero-and-posthumous-pulitzer-prize-winner/.

21. Arlisha Norwood, "Ida B. Wells-Barnett," National Women's History Museum, 2017, www.womenshistory.org/education-resources/biographies/ida-wells-barnett.

22. Jacqueline Jones Royster, ed., *Southern Horrors and Other Writings: The Anti-Lynching Campaign of Ida B. Wells, 1892–1900* (Boston: Bedford, 2016), 10–11.

23. Tera W. Hunter, *To 'Joy My Freedom: Southern Black Women's Lives and Labors after the Civil War* (Cambridge, MA: Harvard University Press, 1997), 70–72.

24. Robin D. G. Kelley, *Freedom Dreams: The Black Radical Imagination* (Boston: Beacon Press, 2002), 41.

25. Omar H. Ali, *In the Lion's Mouth: Black Populism in the New South, 1886–1900* (Oxford: University Press of Mississippi, 2010), 48–50.

26. See Susan Spellman, *Cornering the Market: Independent Grocers and Innovation in American Small Business* (Oxford: Oxford University Press, 2016); and Tracey Deutsch, *Building a Housewife's Paradise: Gender, Politics, and American Grocery Stores in the Twentieth Century* (Chapel Hill: University of North Carolina Press, 2010).

27. "Piggly Wiggly Historical Marker," The Grocery Stores of Clarence Saunders, *Dig Memphis*, The Digital Archive of the Memphis Public Library & Information Center, https://memphislibrary.contentdm.oclc.org/digital/collection/p16108coll17/id/203/.

28. Paula Giddings, *Ida: A Sword among Lions; Ida B. Wells and the Campaign against Lynching* (New York: HarperCollins, 2009), 175.

29. Leah Penniman, *Farming While Black: Soul Fire Farm's Practical Guide to Liberation on the Land* (White River Junction, VT: Chelsea Green Publishing, 2018), 4.

30. See Beverly G. Bond and Janann Sherman, *Memphis in Black and White* (Mount Pleasant, SC: Arcadia Publishing, 2003), 66–72.

31. Zandria F. Robinson, "After Stax," in *Unseen Light: Black Struggles for Freedom in Memphis, Tennessee*, ed. Aram Goudsouzian and Charles W. McKinney Jr. (Lexington: University Press of Kentucky, 2018), 354.

32. Giddings, *Ida: A Sword among Lions*, 175.

33. Damon Mitchell, "The People's Grocery Lynching, Memphis, Tennessee" *JSTOR Daily*, January 24, 2018, https://daily.jstor.org/peoples-grocery-lynching/.

34. The wave of lynchings across the United States in 1919 would be called the "Red Summer." For more on the Red Summer, see Eve Ewing, *1919: Poems* (Chicago: Haymarket Press, 2019), and Simon Balto, *Occupied Territory: Policing Black Chicago from Red Summer to Black Power* (Chapel Hill: University of North Carolina Press, 2019). See also Frederick Douglass Opie, *Hog & Hominy: Soul Food from Africa to America* (New York: Columbia University Press, 2010), 54.

35. Ida B. Wells in Alfreda M. Duster, ed., *Crusade for Justice: The Autobiography of Ida B. Wells* (Chicago: University of Chicago Press), 64.

36. Robinson, "After Stax," 354.

37. Lucy E. Parsons, "Southern Lynchings," *Freedom*, April 1892.

38. On May 4, 2020, Ida B. Wells-Barnett was posthumously awarded the Pulitzer Prize in Journalism. See "Ida B. Wells Honored with Posthumous Pulitzer," Equal Justice Initiative, May 4, 2020, https://eji.org/news/ida-b-wells-honored-with-posthumous-pulitzer/.

39. Monica White in Brian Hamilton, "Food Is Just the Beginning: A Conversation with Monica White," *Edge Effects*, July 16, 2019, https://edgeeffects.net/monica-white/.

9

Modern Food as Racialized Performance

BLACKNESS AND BANANAS
THE JOSEPHINE BAKER EFFECT

Tashima Thomas

In 1927, Josephine Baker performed a dance in a banana skirt at the Parisian music hall Folies Bergère. Recorded as a film short and played before movies, Baker's banana skirt performance would come to influence artists, architects, dancers, Carmen Miranda's famous headdress, and a United Fruit Company (UFC) ad campaign to sell its Chiquita brand bananas. As an African American woman from St. Louis, Missouri, who settled in Paris, dancing in a banana skirt conjured long-standing associations with bananas, Blackness, and primal sexuality. The relationship between Blackness and bananas came to the fore in the nineteenth and early twentieth centuries, an association that connected trusted brands to consumers based on questionable racial and gendered assumptions. Prior to the 1880s bananas were a luxury food item, not unlike the initial production of sugarcane centuries earlier. Their accessibility was limited to exotic varieties that proved expensive and were reserved for special occasions. As Baker danced in Paris, the UFC had grown to become a $100 million company; in the decades to follow it would only grow more. Due to this popularity and the financial success of the UFC, the racialized images of bananas came to be as important as the actual product—moving from the consumption of bananas to the consumption of images. Many of these images were rooted in racial codes of inferiority, a relationship that has become part of our everyday culture.

Baker's iconic performance incorporated a number of key ingredients. In the film short her banana dance is set under the canopy of a jungle's dense foliage, underscoring a kind of primitive paradise atmosphere. Baker descends a fallen tree like a staircase from rainforest heaven, creeping along to meet the base camp of a colonial explorer. The white male explorer attired in a late nineteenth-century hunting outfit complete with pith helmet is retiring for a nap in his tent. He is surrounded by at least five Black natives who are only clothed in short briefs as they attend to their charge. They are styled in the manner of minstrelsy popular at the time, painted in extra dark makeup, emphasizing oversized lips.[1] Baker reaches the floor of the base camp and slides to center stage, flawlessly landing splits with her legs splayed and the skirt of bananas dangling betwixt. She rises up in a bikini top, twirling and rotating her hips in a circular fashion, sending the bananas beneath into graceful flight. She extends her arms overhead, hands meeting, and wrists twisting to the rhythm of the fiery jazz horns. She also draws upon her vaudevillian roots by incorporating comedic movements into her dance, such as bending over and moving her legs inside and out like the wings of a butterfly while crossing her hands upon her knees. She flows effortlessly from the slapstick to the seductive. The steady tempo of pelvic thrusts is directed at the supine colonialist as if sprinkling his dreams with fantasies of erotic desires. Baker is the projection of his colonial reverie.

The banana skirt is believed to have appeared to Surrealist Jean Cocteau in a dream and would soon thereafter be crafted by artist Paul Colin. Then in 1927, the skirt transformed into an abstracted, bedazzled banana motif under the direction of Baker's own costuming design. Baker would redesign the banana skirt as various phallic iterations throughout her career, with her routine interlaced among a range of cultural productions of the era (figure 9.1).

The magnetic intensity of her *danse sauvage* contributed to Baker's rise and stimulated the audience's imagination as they called for more performances. The *danse sauvage* was a pas de deux, a duet performed with dancer Joe Alex. Baker's syncopation and fiery rhythms pulsated throughout her body, moving her backside to the music while incorporating comedic and animalistic postures. Theatre producer André Daven

9.1   Lucien Walery, *Josephine Baker in Banana Skirt from the Folies Bergère production "Un Vent de Folie,"* Paris, 1926. (Photo courtesy of Niday Picture Library/Alamy Stock Photo.)

had added an "African" dance to the show that was choreographed by Jacques-Charles. Daven recalls Baker's performance:

As she danced, quivering with intensity, the entire room felt the raw force of her passion, the excitement of her rhythm. She was eroticism personified. The simplicity of her emotions, her savage grace, were deeply moving. She laughed, she cried, then from her supple throat came a song, clear at first, then with a hoarseness that caught at the heart.[2]

This recollection of Baker's performance helps us envision her dance as it connected with the audience, even before she donned the banana skirt. In the context of Paris's Roaring Twenties, Baker represents the appetite of the colonial project for Black female bodies subjugated and served as consumptive entities. The bananas emphasize the palate seasoned for "Eating the Other," which, in bell hooks's interpretation, denotes the colonial desire to consume the Other as seen through visual and material culture, a way to incorporate "spice" into colonial life.[3] For me, Eating the Other helps us consider the visual consumption of Blackness through commodification. The bananas themselves, a commodified fruit, serve as coded symbols of Blackness. The erotic intensity of Baker's banana dance readily signals sexual consumption where the bananas-as-phallus inspired such associations. Taken together, gustatorial appetites and sexual appetites are intertwined among the ingredients of sexuality, bananas, comedy, and, of course, Blackness. The skirt was indeed more than just a simple fashion choice. Josephine Baker, her banana skirt, and its subsequent connection to the dominant pseudo-colonial company, the UFC, provide a view of how racially coded performances encouraged a new way to trust foods and brands—in this case, bananas and fruit.

RECIPES OF COLONIAL DESIRE

Bananas have long been viewed as derogatory analogs of the Black subject. Dating back to nineteenth-century race science, bananas have been used as political markers of Black identity, associated with primates and reflections of nascent Black sexuality. Race science was promoted by physicians and scientists who advanced notions of white supremacy and biological differences, namely theories of Black inferiority, according to pseudo-scientific discoveries. As critical theorist Homi Bhabha points out, the repetition of racial discourse such as that surrounding the banana creates cultural mythology. "The *same old* stories of the Negro's animality, the Coolie's inscrutability or the stupidity of the Irish *must* be told (compulsively) again and afresh," Bhabha writes, "and are differently gratifying and terrifying each time."[4] Because race science associated Black people with primates and since one of the food sources associated with primates is bananas, Blackness was also associated with bananas, as well as with primal sexuality in a way that became formulaic, a recipe. Historians Steve Striffler and Mark Moberg speak to these ingredients and their attributes:

From the early nineteenth century to the late twentieth, U.S. popular discourses have linked the fruit to dark, sensuous, bumbling, lazy, non-English-speaking people. The enduring use of the term *banana republic*—which first appeared in an overtly racist text—suggests that the banana continues to be a metaphor deployed in order to distance the United States from the tropical neighbors with whom it has been intimately connected for a century.[5]

This particular racial discourse of Blackness and bananas becomes a recipe for a visual construction that girds the banana skirt within a long-standing mythology. Recipes are seldom performed only once. Rather, recipes are repeated over and over again, "compulsively, again and afresh," sometimes perfecting the desired outcome, or modifying according to taste. In Baker's case, the metaphor of recipe-as-performance collided with an actual culinary recipe.

CUSTARD JOSEPHINE BAKER

Alice B. Toklas, partner of Gertrude Stein and part of a tight community of 1920s culture and celebrity, wrote a recipe for "Custard Josephine Baker." It starts with eggs, sugar, milk, kirsch, and the liqueur Raspail, and it ends with bananas. The recipe was likely inspired by Baker's Parisian performance at the Folies Bergère, where she first danced in a banana skirt. Toklas advises to serve the custard cold, although Baker herself served her performance red hot. Here, Baker is the one who is baked and served as a creamy banana custard. Toklas's *Cook Book* is a food memoir of her life with partner Gertrude Stein and the life they shared with many talented and renowned friends and acquaintances. Several other recipes are named after their famous friends.

As a story about Baker, bananas, and modern performance, the recipe was a bond that held together Baker's place in culture with the commodity around her waist.

Toklas's recipes, even the fallible ones, function not just as text-based instructions, but also as scripts for players in a performance. Her cookbook was published with untested recipes that may have even ruined dinner parties over the years.[6] And yet recipes are not merely about the end product. Even the root of *recipe*—the Latin *recipere*—implies an exchange, a giver and a receiver. Far from being an applied science of formulas, Toklas's recipes are akin to sketches of memory that become her personal truth. Other essays in this book touch on modes of trust-building that marked a new era in food production and consumption, through brands (chapter 7), endorsements (chapter 10), and the avoidance of deception (chapter 6). Baker's dance, as a recipe performed, garnered trust by creating alluring associations with consumers, by providing suggestive movements that made the banana appealing.

Baker was not alone in what would become a way of performing a dinner of white desire. The performativity inherent in recipes is probably most clearly represented in Filippo Tommaso Marinetti's *The Futurist Cookbook*, published in December 1930. After writing his infamous *Futurist Manifesto* in 1909, Marinetti turned to a revolution in experimental gastronomy. The provocative *Futurist Cookbook* banned the consumption of pasta in Italy, the use of knives and forks, and after-dinner speeches, as well as several other conventions. According to Marinetti, futurist cooking was "tuned to high speeds like the motor of a hydroplane," seeming "to some trembling traditionalists both mad and dangerous."[7] Literary critic Sandra M. Gilbert describes *The Futurist Cookbook* as a wholly new art form that reimagines cuisine as a kind of abstraction, repudiating bourgeois culinary traditions, and with "radically innovative, often bizarre meal plans [that] foreshadowed the avant-garde 'molecular gastronomy.'"[8] A kind of gastronomic theatre of the absurd, many Futurist recipes were culinary performances designed to enhance the senses. Italian Futurists were also observant of African American performers in Paris, especially of Josephine Baker, often referring to them as "primitives" and thus reproducing racist ideologies regarding Blackness. In 1925, Italian Futurist Enrico Prampolini saw Baker perform in *La Revue Nègre* and offered her a position in his dance company, which she turned down. One Futurist recipe, titled "a dinner of white desire," is an example of the incorporation of Blackness and the performance of recipes. It reads:

Ten Negroes, each holding a lily in his hand, gather round a table in a city by the sea, overwhelmed by an indefinable emotion that makes them long to conquer the countries of Europe with a mixture of spiritual yearning and erotic desire.

Without a word a Negro woman cook serves them twenty fresh white eggs which have been punctured at both ends to inject the insides with a delicate perfume of acacia flowers: the Negroes inhale the contents of the eggs, without breaking the shells.

The Negro cook returns again with a tray laden with pieces of coconut studded with nugat, enclosed in layers of butter and arranged on a bed of boiled rice and whipped cream. Contemporaneously they drink undiluted anise, grappa or gin.[9]

The recipe defines a group of Black performers and ingredients as edible "white desire." The recipe also yields to a hypnotic banquet of absorbing scents of jasmine, anise, and acacia, coupled with the exacerbation of an indefinable emotion. The performative aspect of the recipe is linked to the ten Negroes "longing to conquer countries of Europe with a mixture of spiritual yearning and erotic desire."

Even if she was not alone in the performance of white desire, Baker conquered the hearts and imaginations of Europe in a similar manner, dancing as part of that larger cultural context of interwar desire and yearning. Her stage presence led to the particularly influential "Baker Effect," the term scholars have used to refer to the powerful influence of Baker throughout artistic, cultural, and political domains. The Baker Effect describes her impact as a phenomenon across time. Baker, by mixing a yearning for social acceptance, racial harmony, and humanitarian fellowship with the performance of erotic desire, became an icon of the modern zeitgeist.

THE BAKER EFFECT

Whether it was the iconic status of a banana skirt, the potency of her celebrity, or some combination of the two, Baker cast an indelible impression upon modern artists and architects in ways that, as we will see below, also elevated the sales of tropical fruit. Her impression upon those artists and architects came first. In 1927, to take one example, Alexander Calder created a wire sculpture of Baker, capturing a moment of the frenetic movement and charm of one of Baker's performances. Later, Swiss-French architect, designer, author, and Modernist pioneer Le Corbusier wrote a ballet for Baker. While on a cruise to South America with the chanteuse, Le Corbusier wore one of her costumes, going so far as to be painted in Blackface. Her influence continued. Henri Matisse created a life-size cutout portrait of Baker that he placed in his bedroom. The Futurist Marinetti "praised Baker's legs as 'afrodisiaci pennelli di color cioccolato al latte' [aphrodisiac brushes the colour of chocolate milk]."[10] In 1928, architect Adolf Loos designed a house for Baker. The two had met a year prior when Baker gave Loos dance lessons. Although the plans for Baker's home would go unrealized, "Adolf Loos's model for Baker's house was a dress rehearsal for modern architecture. Though unbuilt, it was destined to become one of his most famous works by virtue of its innovative juxtaposition of exterior and interior elements."[11] The three-story home drew upon Viennese designs, as well as on African and Mediterranean architecture, and was designed with heightened visibility so that Baker could be viewed from every angle. Accommodating Baker's habit of swimming in the nude, Loos included a "glass-encased swimming pool on the second and third floors" that could hold two tons of water.[12]

Composers wrote songs for Baker that she performed internationally. She posed for premier art and fashion photographers. Sociologist Bennetta Jules-Rosette describes how "her libertine image was used to advertise hair pomade, skin lotion, culinary recipes, cigarette cards, and summer vacations."[13] Automobile manufacturer Voisin gave Baker a touring car that she customized with copper-tone paint and brown snakeskin upholstery, in which she cruised around Paris and even participated in automobile rallies. Baker became the test kitchen recipe for how to harness the marketing power of a pop star. These examples demonstrate the Baker Effect, her influence among her contemporaries, and how she represents a modern aesthetic in art and architecture. But, what about a contemporary Baker Effect? How relevant is Baker's banana skirt performance today? In short, it has held up.

The optical insistence of Baker's banana skirt performance has survived across the century after its original performance. Consider the 2003 Academy Award–nominated film *The Triplettes of Belleville*, in which an early scene introduces characters with striking resemblance to jazz guitarist Django Reinhardt, dancer Fred Astaire, and Josephine Baker herself on stage in an early jazz era cabaret. Baker enters stage left dancing on the balls of her feet. She is nude save the skirt of taut bananas wrung around her hips. The animated Baker enacts some of the comedic styling that recalls Baker's early days. Her areolas and nipples swing in a wobbly fashion mirroring her crossed eyes, allowing for a rhythmic hypnotic effect. A long beaded necklace dangles from her neck to her waist, drawing the eye down to the bananas. The crowd is energized, whistling and whooping its approval. The bare voluptuousness of Baker's body, the comedy, the whirling, the bananas, all prove too much for the scrawny male audience-goers. The petite men seated next to their zaftig female companions transform at the sight of Baker's sensual gyrations. The tuxedoed men evolve into screeching monkeys, leaping onto the stage and ravishing the banana skirt in a primal feeding frenzy. Baker is at once revolted and realizes she is in danger from the little beasts who stand to eat her alive. The scene is terrifying. The aggressive monkeys whip up a dust cloud attacking one another as Baker covers herself with her hands trying to protect herself. She runs offstage, leaving the bearded monkeys behind as fast as she can. The animated medium tempers the severity of the horror. This scene reveals how uncritical representations of Baker are disseminated and how the devouring of Baker is also a reflection of sexual assault and the violence of Eating the Other. Witnessing this scene makes it hard not to be struck by the sensational power of Baker's banana skirt performance almost a century later.

The Triplettes of Belleville proves the durability of the ingredients of Blackness, hypersexuality, comedy, and bananas, a performative recipe that Baker tried to eschew the rest of her extensive career. Despite the countless efforts to refashion her image through sartorial elegance, cross-dressing, as civil rights activist, or as French patriot, Baker was never able to slough the legacy of a banana skirt. The Baker Effect was rampant in the interwar period, but also evident in The Triplettes of Belleville decades later. It would appear most dramatically and have the most durable historical effect for a story about modern food with some ingredient variations in the 1940s through the creation of Chiquita Banana. To stick with the metaphor of performance as recipe, Baker's banana skirt recipe was modified according to the tastes of the UFC as its marketing ambitions were ready to take center stage.

THE UNITED FRUIT COMPANY AND CHIQUITA BANANA

Bananas were long considered a luxury food item in the nineteenth century due to their scarcity in marketplaces. But bananas were also popular among Black sharecroppers and former slaves who cultivated the fruit in their own gardens, which deepened the association between bananas and Blackness. Moving bananas into newly forming middle-class homes became the goal for early twentieth-century industry marketing campaigns. The UFC began in 1899 after its successful business ventures procuring bananas in Jamaica and selling them for a profit in New Jersey. By the time Baker took the stage at the Folies Bergère in Paris in 1927 the UFC had come to be worth over $100 million, employing 67,000 people and owning 1.6 million acres of land.[14]

From bananas, UFC built an empire. In a well-told story, the small states of Central America had come to be known as the "banana republics." UFC's reach extended to Belize—former British Honduras—and to Caribbean islands such as Jamaica and Cuba. In South America, Colombia and Ecuador had come under its sway. A company more powerful than many nation-states, it was a law unto itself and accustomed to regarding the republics as its private fiefdom.[15] As I have written elsewhere, the consumer popularity of the banana was directly related to the rise of the banana republic and the international success of the UFC, which formed a close bond with the burgeoning imperialist reach of the United States by the turn of the twentieth century.[16]

With competition on the rise, the invention of Chiquita Banana evolved as a way to distinguish UFC bananas from their competitors Standard Fruit and smaller companies. As Anna Zeide explores in her essay on canned food (chapter 10), modern food producers often felt the need for celebrity endorsements to sell their goods. At times companies fashioned those celebrities themselves, modeled on real performers. In 1944 the "Miss Chiquita" banana persona was created by cartoon artist Richard Arthur "Dik" Browne (1917–1989). Browne was best known for his creation of the comic strips "Hagar the Horrible" and "Hi and Lois." The UFC invited him to create an animated banana that would serve as a brand ambassador for Chiquita. The image was inspired by Brazilian sensation Carmen Miranda's performance in the 1943 Busby Berkeley musical *The Gang's All Here* (figure 9.2).

9.2 Carmen Miranda in Busby Berkeley's musical *The Gang's All Here*, 1943. (Image courtesy of 20th Century Studios, Inc.)

In the film, Miranda's bananas and strawberries number manifests food's gender assignments. The phallocentric ubiquitous bananas share the stage with the ripe strawberries representing fecundity and male sexual desire.

Browne's incarnation of Chiquita Banana blossomed into a series of animated film shorts. Capitalizing on the fruits of a consumer survey showing that households with children ate more bananas, the company targeted children as current and future customers, including, as historian Virginia Scott Jenkins writes, sending "materials to classroom teachers, including nutrition information, recipes, films, pictures to color, sheet music, maps, and geography lessons."[17] The popular practice of feeding mashed-up bananas to babies and adding bananas to breakfast cornflake cereals that we take as a given today were linked to the marketing advertisements of the UFC. In addition to pandering to children, the animated films could also soften the reputation of a banana magnate who had deposed foreign leaders, established puppet governments, and led military coups. As a response to its conflicts with US antitrust laws, the UFC had consistently reorganized and rebranded from UFC to United Brands to Chiquita Brands.[18]

Within this context of military aggression and legal action, Miss Chiquita appeared as a spokesperson with a fruit-laden hat celebrating the nutritional value and preparation of bananas. She appeared to midcentury consumers as yet another instance of the Baker Effect. Miss Chiquita starred in thirteen short animated films produced by Famous Studios in the late 1940s including *Chiquita Banana, Chiquita Banana Goes North, Chiquita Banana Helps the Pieman,* and others. The film most aesthetically derivative of Baker's banana skirt dance was *Chiquita Banana and the Cannibals.* The cast of characters is quite similar to Baker's 1927 film. A white male explorer in hunting outfit and pith helmet sits next to the minstrelsy African native with dark skin and oversized lips. But a reversal of fortunes has taken place. Rather than pampering the colonialist settling down for a nap, the explorer is being cooked in a large iron pot and seasoned to taste by the African native. The opening scene plays upon the popular trope of explorers encountering cannibal natives layered over a stereotypical African drum beat commonly orchestrated in early Tarzan films. The tropical-themed music then begins as both characters stop to watch Chiquita Banana swaying her banana body seductively to the rhythm. They are captured by the exoticism and erotic appeal as Chiquita sings:

I'm Chiquita banana and I've come to say
That you really shouldn't treat a fellow man this way
If you'd like to be refined and civilized
Your eating habits really ought to be revised

The African native in this scenario is playing the role of the *unrefined* and *uncivilized*. After saving the colonialist from the cooking pot, Chiquita advises the two on how to make "Banana Scallops," which is simply a variation of fried plantains. She encourages Eating the Other, which is herself in this example, an exotic entertainment.

The entire scene was an aesthetic, performative, and cultural descendant of Baker's banana skirt performance, which itself had derived from the full flush of racial codes of desire and appeal from the nineteenth century. Closer at hand—in between the Folies Bergère and *Chiquita Banana and the Cannibals*—Chiquita Banana was aesthetically driven by Miranda's performance in *The Gang's All Here*. Yet, even in that case, Miranda's own performative style was based in colonial conditions as inspired by Afro-Bahian women from Brazil. In her case, Miranda specifically modeled her outfits after the traditional colonial dress of Bahianas (Black women from Bahia in the northern region of Brazil). Bahianas wore long skirts and ruffled blouses, and had baskets on their heads full of fruit to sell in the marketplace.

The 1995 documentary *Bananas Is My Business* explains the origins of Miranda's image saying, "Carmen was our dancing doll; white to disguise the Blackness of the music she grew up with; the Blackness of most of her musician friends; and of the origins of her outfit."[19] With Carmen Miranda as a simulacrum of Black Bahianas, Chiquita Banana was essentially based upon the stylings of Black women in northern Brazil who unwittingly or not reinforced the banana skirt recipe of Blackness, primal sexuality, comedy, and bananas. Holding bananas and racial subjectivity together was again a case of the Baker Effect.

In what other ways does the Chiquita story reflect the Baker Effect? At first the Chiquita brand logo was placed on a bunch of bananas that were held together by a band. The trademark blue sticker was affixed to bananas in 1963 as part of a blockbuster marketing campaign with the motto, "This seal outside means the best inside."[20] It was not the first time a sticker was placed on bananas as part of a marketing campaign. Baker's 1934 film *Zouzou* was written by her partner Giuseppe "Pepito" Abatino and featured a Cinderella story that ends in an unrequited love triangle. In his efforts to promote the film Pepito created labels for bananas that read "Josephine Baker is Zouzou." As Jules-Rosette writes in her Baker biography, Pepito "encouraged shopkeepers throughout Paris to display these bananas, and he also developed an internationally distributed Zouzou newsletter containing advice on love, fashion, and daily life."[21] Pepito drew upon the legacy of the banana skirt performance by conflating bunches of bananas with Baker's image. The sophistication of Baker's and Pepito's marketing prowess was a model not just for performers, but also for banana companies. In the 1945 Chiquita technicolor film advertisement the sashaying Chiquita sings, "You'll find by eating fruit you'll have a more beautiful appearance and complexion," and "a daily dose of bananas will help you look perfection."[22] Even beyond that instance, it was Baker's advice on beauty and marketed beauty products such as

"Bakerskin" that entered mainstream success. Bakerfix too was a brilliantine cream created by an Argentine chemist. Billboards and posters advertising Baker's beauty products were seen all around Paris.

There was one more way in which later UFC branding carried forward the Baker Effect from the 1920s; this was the name Chiquita itself. Baker famously had a pet snake named Kiki and a pet cheetah named Chiquita, and long before UFC used the name Chiquita for their brand of bananas Baker often strolled the promenades of Paris with her pet cheetah of the same name. The publicity machine of Pepito and Baker popularized this exoticism, and the graphic artist Zig used Chiquita the cheetah as the model for Baker's *Casino de Paris* poster of 1930 (figure 9.3).

The aesthetic and historical correlations between Baker and Chiquita Banana draw attention to the Baker Effect as an archetypal model of Blackness, eroticism, sexuality, comedy, and bananas. One might not consider Baker and Chiquita Banana's affinities obvious, as their bare names alone do not suggest the connections. Yet, rather than mere coincidence, the UFC's marketing strategies were intentional business practices fully informed by the Baker Effect. Applying stickers to their bananas, associating bananas with beauty and health regimens, emulating Baker's filmic performance in *Chiquita Banana and the Cannibals*, the visual aesthetic of an exoticized dancing banana, and rebranding itself after the name of Baker's pet cheetah all derived from Baker and demonstrate the modern marketing of bananas. Through the altering of certain ingredients, the recipe for Baker's visual aesthetic had become part of the Chiquita Banana persona. The marketing prowess behind bananas throughout the twentieth century is a product of the elevation of the banana into the public consciousness as desirable. Consumers came to trust the fruits once dangling from Baker's skirt. But, without the performativity of Chiquita Banana dancing, singing, and advising on recipes and beauty regimens, the recipe for the success of the banana was inactive. Some recipes require a performance.

CONCLUSION

Baker would come to work as a humanitarian and civil rights activist who also fought to relieve the sufferings of the poor.[23] Baker received the Legion

9.3 Zig, *Josephine Baker: Casino de Paris*, 1930. (Image courtesy of World History Archive/Alamy Stock Photo.)

of Honor medal at the Chateau des Milandes in 1961. She endangered her life as a secret agent working for the Counterespionage Services and the French Resistance in France, North Africa, and Spain. She obtained classified information and received the Croix de Guerre and the Rosette de la Résistance for her efforts. She supported the Civil Rights Movement of the 1950s and 1960s, working with the NAACP, speaking at historically Black colleges, and most notably at the March on Washington in 1963 where she stood alongside Dr. Martin Luther King Jr. Her biographer Jules-Rosette describes her role as "an ambassador for racial and moral tolerance [who] made political statements on stage and in the press. These responses framed her political persona and awakened her awareness that a song or dance was not merely a performance but also a political statement."[24] Yet, despite her wartime heroics, championing of the poor, and adoption of twelve children from around the world to prove that racial harmony was possible, Baker is most often remembered for her banana skirt.

Baker had long outgrown the Josephine of the 1920s and the banana skirt, but never seemed to be able to abandon its problematic image.[25] As part of a marketing campaign, the banana skirt and dancing bananas were performative recipes rooted in racial degradation. I came upon Baker's story as an art and food historian, believing that we are not only what we eat, but what we consume—and we consume visual constructions as much as we consume the banana. Our consumption of images becomes our constitution and deserves our thoughtful attention; those images matter for generating appeal but also for embedding and promoting racialized cultural codes for those in the future. The root of *recipe* suggests an exchange, a giver and a receiver, all of which are appropriate regarding the influences of a banana skirt on the marketing of modern food. The Black body, hypersexuality, comedy, bananas, and performance are ingredients that have been well sifted over time and combined to reinforce ideologies of consumption.

NOTES

1. Bennetta Jules-Rosette, *Josephine Baker in Art and Life: The Icon and the Image* (Chicago: University of Illinois Press, 2007). Jules-Rosette describes minstrelsy as "tragicomic performance based on the hegemonic dialectic of race relations" (56).

2. Jules-Rosette, *Josephine Baker in Art and Life*, 61.

3. bell hooks, "Eating the Other: Desire and Resistance," in *Black Looks: Race and Representation* (Boston: South End Press, 1992), 21–39.

4. Homi K. Bhabha, *The Location of Culture* (London: Routledge Classics, 2004), 111.

5. Steve Striffler and Mark Moberg, eds., *Banana Wars: Power, Production, and History in the Americas* (Durham, NC: Duke University Press, 2003), 77.

6. Alice B. Toklas. *The Alice B. Toklas Cook Book* (New York: Harper, 1954), 117.

7. Filippo Tommaso Marinetti, *The Futurist Cookbook* (London: Penguin Books, 1932), 11.

8. Sandra M. Gilbert. *The Culinary Imagination: From Myth to Modernity* (New York: W. W. Norton, 2014), 5.

9. Marinetti, *The Futurist Cookbook*, 186. Several other Futurist recipes require a great deal of performance, including "geographic dinner," "economical dinner," "nocturnal love feast," "summer luncheon for painters and sculptors," and "extremist banquet," where the dinner guests are not allowed to eat food but whose satiety comes from breathing various perfumes as they are surrounded by food sculptures equipped with vaporizers emitting scents of vanilla, red pepper, chocolate, and acacia flowers.

10. Przemysław Strożek, "Futurist Responses to African American Culture," in *Afromodernisms: Paris, Harlem, Haiti, and the Avant-Garde*, ed. Fionnghuala Sweeney and Kate Marsh (Edinburgh: Edinburgh University Press, 2013), 49.

11. Jules-Rosette, *Josephine Baker in Art and Life*, 151.

12. Jules-Rosette, *Josephine Baker in Art and Life*, 145.

13. Jules-Rosette, *Josephine Baker in Art and Life*, 146.

14. Dan Koeppel, *Banana: The Fate of the Fruit That Changed the World* (New York: Plume, 2009), 75.

15. Peter Chapman, *Bananas: How the United Fruit Company Shaped the World* (Edinburgh: Canongate, 2007). Also see the article by Annie Mendoza and Tashima Thomas, "Literary and Visual Rememory at the 90th Anniversary of the Banana Massacre in Colombia," special issue, "Food Fights: A Global Perspective," *Zapruder World: An International Journal for the History of Social Conflict* 5 (Fall 2019). "The consumer popularity of the banana is directly related to the rise of the "banana republic" and the international success of the UFC, which formed a close bond with the burgeoning imperialist reach of the US by the turn of the twentieth century."

16. Mendoza and Thomas, "Literary and Visual Rememory." See Chapman, *Bananas*, for an entry into the fuller story of "banana republics," and John Soluri, *Banana Cultures: Agriculture, Consumption, and Environmental Change in Honduras and the United States* (Austin: University of Texas Press, 2006), for broader context.

17. Virginia Scott Jenkins, *Bananas: An American History* (Washington, DC: Smithsonian Institution Press, 2000), 69.

18. Koeppel, *Bananas*, 143.

19. Helena Solberg, director, *Carmen Miranda: Bananas Is My Business*, filmstrip, 1995.

20. "Best Banana," https://www.chiquita.com/the-best-tasting-highest-quality-banana/.

21. Jules-Rosette, *Josephine Baker in Art and Life*, 79.

22. Jenkins, *Bananas*, 71.

23. Dora B. Weiner. "Francois-Vincent Raspail: Doctor and Champion of the Poor," *French Historical Studies* 1, no. 2 (1959): 149–171.

24. Jules-Rosette, *Josephine Baker in Art and Life*, 214.

25. Jules-Rosette, *Josephine Baker in Art and Life*, 153.

10

Modern Food as Endorsed Food

MARION HARLAND, TASTEMAKER
HOW ONE WOMAN'S INFLUENCE HELPED BUILD AN INDUSTRY

Anna Zeide

In halls and auditoriums across nineteenth-century America, there she was. She stood at the lectern, in a stiff brocade dress and thickly starched white collar, projecting power. She rose before her rapt audiences on the lecture circuit, the women with their brimmed hats and feather plumes, and poured forth her wisdom. The women came because they had read one of Marion Harland's many famous cookbooks or domestic advice manuals, or maybe one of her novels. Her *Common Sense in the Household: A Manual of Practical Housewifery* had been published in 1872 and had sold over a million copies. Women gave dog-eared copies of it to their newly married daughters, urging them to turn to Harland when they needed to take care of sick family members or restore the pile of a velvet dress, shine silver or remove berry stains, make fritters and brandied fruits, organize a pantry, and host a party of twelve. Or perhaps the women in the audience had turned to Harland's 1885 *Common Sense in the Nursery* in their early days of tending to a crying infant in the wee hours of the night, searching for guidance and finding solace in Harland's confident advice.

Marion Harland, the pen name of Mary Virginia Terhune, was, according to her biographer, "the Julia Child, Danielle Steel, and Dear Abby of her day."[1] She was a well-known writer and a popular figure on the Chautauqua lecture circuit in the 1890s, an educational institution that brought speakers, preachers, and entertainers to rural America in the late nineteenth and early twentieth centuries.[2] In addition to over seventy books, she also published hundreds of articles and stories and had syndicated newspaper columns in the Philadelphia *North American* and the *Chicago Tribune*. Through all of these venues Harland spread her gospel of domestic expertise far and wide and connected with adoring fans.

Marion Harland's influence did not go unnoticed by those who sought the attention of American housewives in the late nineteenth century. The canning industry emerged during this time on first-shaky legs, ushering in an industrialized approach to food in America. We now take its success for granted, assuming its ubiquity. Canned foods are everywhere; before Harland's time they weren't. At the height of Harland's fame canned food was still unknown, unsavory, and untrustworthy. In order to become more sure-footed, the food processors needed to win over American consumers. They needed a spokesperson who carried authority and trust when it came to matters of the kitchen—a spokesperson like Marion Harland.

By 1910 the canning industry had gotten her on board, commissioning Harland to write a cookbook published by the National Canners Association called *The Story of Canning*. But that wasn't an obvious fit. Just over a decade earlier Harland had published derisive takedowns of canned foods, warning her followers away from them. So, what happened? How did such a stark transformation take place in the slim intervening years, for Harland to go from hating canned foods to loving them?

This story of Harland's change of heart takes us into a story of the food industry and home economics, into the launch of appeals to celebrity, product placement, and heavy emphasis on advertising that are today such staples of the modern food industry.

The fifty-year span between 1872 and 1922 in which Harland was writing books on cooking and homemaking is nearly the same half century during which food production in the United States turned toward its modern form: dependent on processed foods and on scientific understandings of nutrition. Celebrities helped assure this change. Akin to Tashima Thomas's description of Josephine Baker's influence on the interpretation of bananas (chapter 9), in this story a prominent white celebrity helped fashion a different reception for canned food. A key strand in modernizing the food supply through canned products was the securing of Harland's influence.

In the late nineteenth century the canning industry was still in its rela-tive infancy, growing slowly and meeting much consumer resistance. The process of canning involved preservation of food by heating in a vacuum-sealed container. The credit for its invention goes to Frenchman Nicolas Appert, who developed the process around 1800 in response to a call from Napoleon to find a way to better feed his army. Canning slowly spread from there to England and then to the United States. The canning industry gained a foothold after the Civil War, seeing a pathway beyond the existing market of sailors, soldiers, and specialists and into ordinary consumers' homes. But by 1896 the industry was still bringing a product to the American table that most consumers considered distasteful and suspect.[3]

The men of the canning industry—for they were mostly men—understood the uphill battle they had to fight in improving not only the technology of canning but also its image. The National Canners Association (NCA) was officially founded in 1907, on the heels of the 1906 passage of the Pure Food and Drug Act, for which the leading canners had levied their support. In organizing as a national body, canners across many different kinds of foods saw strength in their ability to build an industry through both tan-gible technical improvements and image promotion. Even as they worked with technologists, bacteriologists, and agricultural scientists to improve the food inside the can, they also turned their attention toward convincing consumers that these foods were worth buying.

The problem was, not only were consumers skeptical, but they were also confronted with negative attention directed toward canned food around the turn of the twentieth century. Spoiled canned food was often blamed for food poisoning. Media reports maligned the products. One 1907 *New York Times* headline read, "Canned Peaches Fatal: Two Boys Die from Pto-maine Poisoning—Mother May Die."[4] And famous writers like Marion Har-land were initially among the industry's biggest critics.

In the late nineteenth century, as she shared her opinion of canned foods, Marion Harland did not mince words: "the very mention of canned goods," she wrote, "is productive of a disgustful qualm." In her view, canned foods were "insipid products of factories" and "plebeian article[s]" that her readers should judiciously refuse.[5] There was little place for these foods in Marion Harland's kitchen. Her daughter, Christine Terhune Herrick, with whom Harland worked closely, similarly rejected the industrial product. Expressing her own disgust, she described her experience of opening a can of chicken only to find "the whole head—beak, wattles, and comb—of a rooster." Grossed out, Herrick decided "with a sick shudder . . . to renounce canned foods and all their works."[6] Marion Harland and her daughter shared their own bad experiences with canned foods to convince their readers to stay away.

But let's jump forward fifteen years. In 1910 we find Marion Harland putting forth a very different set of views. No longer an affront to the kitchen counter, those "insipid products" now appeared in Harland's recipes. Harland had even written an entire book published by the NCA called *The Story of Canning*, full of dishes reliant on canned food. In the book's introduction she described how a "sense of justice" and a "desire to help [her] fellow-housewife" had driven her to write the cookbook and promote these canned foods that she acknowledged she had previously considered taboo.[7] She devoted the first portion of the book to an exploration of the scientific principles that underlay the canning process and to the purity of the industry—all offered as explicit justification for her change of heart. Harland then offered 9 recipes for soup, 13 for fish, 29 for vegetables, 10 for salads, 62 for desserts, and 10 for meat—every single one beginning with a version of "open a can of . . ."

Following in her mother's footsteps, Christine Terhune Herrick published her own pro-industry about-face in 1914 in the *Woman's Home Companion*. She described the experience of grudgingly trying a can of peas that her grocer had convinced her to buy and finding them so surprisingly delicious that, "with the zeal of a new convert, I longed to advocate the use of canned foods." She wrote in romantic terms about the sanitary process that brought peas from field to can, the great machines that peeled and washed and sorted the vegetables through a controlled, methodical, modern system.[8]

It would be easy to simply write this off as a straightforward story of selling out. And almost certainly the canners paid at least Harland for her efforts in writing the book that was published under their auspices. But by the time she wrote the book in 1910 she was eighty years old and looking back on a long, successful, lucrative career. She was still a powerful figure, working up to eight hours a day. Her son Bert wrote that up to her death at age ninety-one she was "full of strong commonsense and imagination. Her mind blazed clear and glowingly optimistic."[9] She had always been a woman of integrity and singular purpose, committed to her principles. This wasn't just about money. So, although she accepted the canners' payment, what else allowed her to take this compensation while maintaining her integrity? Diving deep into this question takes us on a journey into industrial food production, domestic science, and the increasing power of the food industry. Marion Harland's period of conversion captures the moment in American history when the question "What do I eat?" began to be accompanied by a parallel question: "What do the food celebrities eat?" These food icons were born from and endorsed by a partnership with industrial food manufacturers, who had distinct interests in what modern American consumers ate.

To understand this apparent sea change, let's first take Harland and Herrick and their word. What were their own explanations for the conspicuous changes of heart?

Christine Terhune Herrick's stated motives were relatively straightforward: she pointed to the technical improvements in the canned products themselves. She described how quickly peas were brought from field to factory—sometimes in just four hours—where they were handled by a series of elaborate, sanitary machines. This kept canned foods in "perfect cleanliness," which couldn't be attained by the "housekeeper who struggles to keep her kitchen and its contents free from dirt, germs, and consequent disease."[10] Writing during a time when an awareness of the germ theory of disease was spreading from the laboratory into the American home, Herrick's reference to cleanliness, germs, and sanitized machinery would have rung true with many American women similarly enticed by the "natural" processed foods described in Michael Kideckel's essay (chapter 7), who had begun to worry about what kinds of invisible foes lurked in the corners of their kitchens.[11]

Herrick raved about the "freshness" and quality of these canned peas. She swore she'd choose them over the "withered and yellow supplies" she could purchase at the market from a greengrocer. She argued that the price of canned peas had fallen not because of reduced quality but because of technological improvements. Finally, she attested that the Pure Food and Drugs Act, passed in 1906, kept canned food free of chemicals, preventing the "unscrupulous man" from covering up the taste of rotten vegetables with chemical additives. All of these improvements were a sign, in Herrick's view, of a steady upward trend in the American quality of life, a classic view of the Progressive Era.[12] Leaving no exaggeration unprobed, she wrote, "The ability to secure good canned foods should do much to . . . make life worth living."[13]

Harland, on the other hand, went a bit loftier in her explanation. Harland began *The Story of Canning* with an exposition on the way that the traditional "methods of protracting the usefulness of the most precious products Mother Earth offers her children"—that is, of preserving fruits and vegetables—were "Desiccation and Salt." She lamented the fact that these methods destroyed the "life and essence of all green and growing things—SUCCULENCE . . . [which] regulates the biliary secretions and the action of the digestive organs, and purifies the blood." Harland described how these old ways were being supplanted by the new wave of scientific knowledge that birthed the great canning method. Even with this long history of preserving food, and even with the acceptance of sterilization and germ theory in modern hospitals that she described, Harland was aghast. She burst off the page in frustration that the consumer of 1910 still couldn't accept canned food. "Yet," she wrote, "and herein is mystery!— the methods of the conscientious canner sound to us like fairy tales and exaggerations ludicrously incredible." She attested to the sanitary process, the gentle handling, the compliance with pure food laws that she learned about when she "entered upon a careful study of this subject."[14] Without this careful study, or the vote of confidence by a domestic expert such as herself, she understood that most readers might have no idea how much the industry had improved and now deserved their patronage.

Some of these claims may have seemed overblown, but they were not ridiculous. Indeed, canned foods *had* markedly improved between the 1890s and 1910s and would continue to improve further in the decades to come. The canning industry was hard at work, collaborating with a large network of scientists to make their products taste better, last longer, and cost less. Bacteriologists were investigating how time and pressure of cooking affected microbial growth; agronomists and farmers were collaborating to improve the raw crops that ended up in the can; inventors and engineers were developing machines that streamlined the production of tin cans and that more efficiently processed the fruits and vegetables that went into those cans.[15]

The canners understood, however, that it wasn't enough to just make these improvements. No, they had to *communicate* these changes to the broader consuming public. And what better way to do this than through the voice of someone who had standing-room-only crowds at her lectures, whose syndicated newspaper columns were widely read, and who had published many best-selling books of domestic advice? Marion Harland was a woman who commanded attention. If the canners could get Harland, her daughter, and their command of millions of American housewives on board, the lobbyists' dogged advances would not be for naught. They thus set out to woo these prominent influencers. It would be nice to take Harland and Herrick at their word, but in keeping with the coming century of such lobbyist influence over what we eat, the canning industry was more involved than Harland and Herrick may have let on.

So, although both women made it seem that their newfound knowledge of the improved canning process came from a period of self-study, this education was actually intentionally conveyed to them by the canning industry. Herrick's 1914 claims that she "undertook to investigate" the new methods of canning on her own as the result of natural curiosity seem especially disingenuous given that her mother had written a book on these same topics, published by the NCA a full four years earlier. And by this time, Marion Harland, widowed and in her eighties, was living with her daughter, which meant they were likely in very close contact about their mutual work. Herrick explicitly addressed this concern in her essay, denying any outside influence. "I am not writing this talk," she insisted, "with the intention of pushing the interests of the makers of canned foods. I hold no stock in a cannery of any sort (I wish I did!)."[16] And yet, whether or not Herrick's aim was to intentionally push those interests, by 1914 she and her mother had become deeply intertwined in an enterprise that very much served the interests of the canned food makers.

The organization of the National Canners Association in 1907 gave canning men the platform from which to jump in with full force, unwilling to sit back while the influential tastemakers advised their would-be consumers to snub canned foods. They launched an active, intentional campaign to make canned food the darling of home economists by the second decade of the twentieth century.

The NCA formed a publicity bureau in 1909, in order to promote the positive reputation of canned goods. Within a year the bureau was on record as having "presented facts regarding the wholesomeness of canned foods to the Associated Domestic Science Clubs," with a special focus on the audience member they valued most: "particularly to Marian [*sic*] Harland, the leading syndicate writer of food articles of those times." Subsequently, Marion Harland "made public appearances at several early canning conventions," including the 1912 convention in Rochester, New York, braving the "intensely cold" weather to hobnob with the nation's leading canning men.[17]

Harland came to many of her reformed positive ideas about canned food through direct exposure to the men who were making a name for the industry. In *The Story of Canning*, Harland cites "a most interesting, instructive and suggestive lecture delivered at the recent Pure Food Exhibition in New York City, by Mr. Hugh S. Orem, President of the Booth Packing Company of Baltimore."[18] Orem was also a former president of the Canned Goods Exchange, the first canners' trade association, formed in Baltimore in 1883, when the majority of the industry was centered in that city. And he served on the NCA's executive committee from 1908 to 1910.[19] He was a booster, presenting his rosy view of the industry in places of influence. And it paid off.

These interactions between Harland and the canning industry clearly had an effect on both parties. After Harland learned from Orem's "instructive" lecture at the Pure Food Exhibition and after the NCA's publicity bureau convincingly made its case to her, she published her canning cookbook with the NCA in late 1910. *The Story of Canning* drew heavily on the kinds of scientific language that Orem presented in his address, translating technical ideas into narratives more accessible to the average consumer. The NCA, in turn, was so pleased with this early instance of engagement with a prominent domestic expert that the experience went on to shape their agenda in this sphere for decades to come. Indeed, this lay the foundation for the celebrity food endorsements and product placements that are now central to the modern food industry and its advertising. In 1947, for the fortieth birthday of the NCA, industry leaders reflected on their relationship with Marion Harland a half century earlier. "She demonstrated," they wrote, "how it was possible to express the significance of scientific findings in language and terms familiar to the housewife and cook." She was an especially effective translator. This work with Harland made "a profound impression on the trade and for years canners agitated for home economics work as part of Association service."[20] The canners' experience with Harland at the turn of the twentieth century cemented their perception that they had to address consumers in their homes, through trusted public figures, in order to win over consumer confidence.

The burgeoning food industry, and Harland's role in it, was also bolstered by the nascent scientific discipline of home economics. Part of the broader area of domestic science, home economics emerged in the early twentieth century, bringing science from the country's laboratories and factories into kitchens and parlors. Canned foods and other products of the factory embodied this rigorous approach. In the view of domestic science, women were no longer simply *housewives*; they were now responsible for being *household managers*. No longer would their cooking and cleaning be dictated by tradition and intuition—now education, measurement, and analysis would be their guides. Perhaps, some of these domestic scientists hoped, this exacting approach could help transform the American home and, by extension, American society.[21]

Marion Harland rose to prominence in the nineteenth century, largely before the ascent of domestic science. As a result she began as spokesperson for the old guard—the housewife, rather than the household manager. Her early aversion to canned foods was in keeping with this identity. She embarked upon her study of the household arts after her marriage to Protestant minister Edward Payson Terhune in 1856. The newly married couple made their home in Virginia, where Harland forced one enslaved woman to help her with household chores despite her husband's abolitionism.[22] In her autobiography she wrote, "I learned by degrees to regard housewifery as a profession that dignifies her who follows it, and contributes, more than any other calling, to the mental, moral and spiritual sanity of the human race."[23] A profession, yes, but one driven less by a standard body of scientific knowledge than by "common sense in the household"—as her 1871 magnum opus was entitled. Written after the American Civil War, this book sought to educate the many women who had relied on enslaved workers to bear that common sense. And for many northern women, habits and assumptions had transformed through the devastation of war, wartime innovations in food processing, and new connections made possible by the recently completed transcontinental railroad.[24] Harland offered guidance in developing the profession of housewifery.

Common Sense in the Household came out at a time when cookbook publication had increased dramatically, along with promotional literature hawking specific brands of foods and appliances.[25] Even as other cookbooks began to incorporate domestic science, Harland's publications remained rooted in the older style. For example, Fannie Farmer's legendary *Boston Cooking School Cook Book*, published in 1896, included chemical analyses, US Department of Agriculture tables of food composition, and an emphasis on measurement. In contrast, Harland's *Complete Cook Book*, published seven years after, paid little attention to exact measurements.[26] Harland addressed this choice explicitly, writing at the very end of the cookbook: "It is not possible to make out a table which shall be absolutely accurate. Experience is the one trustworthy teacher."[27] The experience of the home was primary in her mind, not the new science or the products with which its practitioners associated.

While this may well have been Harland's self-perception, a few factors swayed Harland's allegiances toward embracing a newer tradition. The first was that she was always quite brand conscious. As branded processed food became entwined with modern scientific management in the early twentieth century, Harland's own perspective on domestic science would shift. Even in her first cookbook in 1871, she promoted a specific brand of desiccated codfish and later endorsed Cleveland's Baking Powder.[28] And, despite her tirades against canned foods in 1896, the promotional materials of at least one brand of canned soup claimed her endorsement several years earlier.[29] Her 1903 *Complete Cookbook* referred to "mother's Impromptu Larder" with one shelf holding "the best brand of canned soups." Throughout the cookbook she allows for the use of some canned foods—lobster, shrimp, mushrooms, and a variety of fruits and vegetables—though nearly always as a second best, if no fresh versions were on hand.[30] All this, even as she disparaged cans in general in her 1896 *National Cookbook*.

Despite Harland's aversion to commercially canned food as a whole, the branded products for which she sometimes made exceptions would become central to the domestic science movement as the values of American business invaded the home in the early twentieth century. Although home economists had first focused on nutrition and hygiene, soon the scientific approach they favored led them directly into the arms of food manufacturers. In the words of writer Laura Shapiro, "Uniformity, sterility, predictability—the values inherent in machine-age cuisine—were always at the heart of scientific cookery, so the era of manufactured and processed food descended upon the domestic-science movement like a millennium."[31] As young, ambitious women graduated from university home economics departments, food businesses stepped in as eager employers, creating an intimate connection between the cooking schools and the industrial giants.[32]

A second major factor in Harland's shift toward industrial food was her daughter. Although Christine Terhune Herrick would also shift from rejecting to exalting canned foods, her commitment to the broader modern food system of which canned goods were a central part was much more steadfast. In 1896, when she and her mother coauthored *The National Cookbook*, there was a distinct shift in tone from the books written solely by Marion Harland. As her biographer writes, "Christine . . . lacked her mother's easy warmth and informality. Christine could tell a reader what to do; she could not seem to join her in the kitchen."[33] Along the same lines, Harland used the term "home-maker" or "housemother" in describing her audience, while Herrick used the more formal "caterer" or "Average American Housewife." These latter terms were often also used by domestic scientists in an attempt to professionalize the home and kitchen. After her mother's death in 1921, Christine Herrick published a final edition of *Common Sense in the Household*, only this time she hired a home economist from Columbia Teachers College to calculate caloric food values and add more scientific content, bridging her mother's older traditional approach and the new scientific focus.[34]

Harland's 1910 embrace of canned foods in *The Story of Canning* shows her on the precipice between the old and the new.[35] She had spent most of her life embodying the older tradition of the housewife, considering experience and the home as the ultimate teachers. During this earlier phase, despite her brand consciousness, she mostly rejected canned foods both for what they were—overcooked, inferior—and for what they symbolized—a new industrial, scientific approach to managing the home. As she grew older, the world around her tilted toward cooking schools, caloric tables, and food manufacturers. She was swept up by this transformation and helped accelerate it. By the time she wrote *The Story of Canning* she had leaned into the scientific approach, using technical and medical language and extolling the virtues of canned foods, those scientific products of industry. In the same year the booklet was published, at the age of eighty, Marion Harland would embrace a new audience, the coming wave of household managers.

In June 1910, Harland gave a speech at the commencement ceremony for the Columbia Teachers College cooking school in New York City. She had previously considered this institution and the profession it promoted to be directly antithetical to the approach she championed. Now, it appeared to her to be an acceptable new direction for homemaking. Maybe the future of the field wouldn't be quite what she had imagined, but by establishing home economics as a respected part of college curricula, she hoped that these kinds of schools would further glorify the profession to which she had devoted her life.[36] Instead, that marriage of domestic advice and processed foods, of home economics and the food industry, helped enable the prepackaged meals and fast food that have relegated cooking and homemaking to the backburner for most Americans.

It took a lot of work to make canning publicly legitimate. The wide array of packaged food that followed it required that acceptability. In the twenty-first century, consumers take that legitimacy mostly for granted. For cans to become so ubiquitous, canned food entrepreneurs a century ago had to overcome ingrained dietary habits and prejudices.

Marion Harland stepped in to fill this need. Her power appealed to the canning men who were just getting their start and who needed the kind of influence she wielded. Despite the strength of the food industry today, back then she wielded power—a great woman among small men. And although she was a woman speaking mostly to an audience of other women, her authority and acceptance of canned food ultimately helped launch the rise of the canning industry, strengthening the power of those men at the helm of the industry.

Her story points to a milestone in modern food choice, helping illustrate a broader point: overarching food systems are the result of many individual choices linked together. When early twentieth-century consumers selected a can of peas from their local grocer, they were likely not thinking explicitly of Marion Harland. And yet, her subtle influence, and the influence behind her of the canning industry and the domestic science movement, was there. Today, your decision or mine, to pick up a can of food from the grocery store shelf may feel isolated, insignificant. But that decision is nestled in the middle of a very long line—before it came the history and culture, advertising and marketing, peer pressure and influence, upon which all consumer choice is based; after it comes the reverberating effects all down the line that build far-reaching consumer acceptance. Our decisions are not ours alone. Taste has to be made.

In the case of canned food, Marion Harland was a central tastemaker. This particular powerful woman—embedded in a moment of scientific, industrial, and professional transition—helped make a place for canned food. To know Harland's story is to know how unfamiliar foods become familiar, how tastes are built.

NOTES

1. Karen Manners Smith, "Marion Harland: The Making of a Household Word" (PhD diss., University of Massachusetts-Amherst, 1990), xi. The biographical details included in this essay are derived from Manners Smiths's excellent dissertation on Harland. That biographical dissertation focuses especially on Harland's literary career and gives less attention to her involvement with food, making no mention of her work with the canning industry, which is what this essay adds to the literature. Karen Manners Smith has been most gracious in helping me think through some of the arguments in my own story.

2. Charlotte Canning, *The Most American Thing in America: Circuit Chautauqua as Performance* (Iowa City: University of Iowa Press, 2005).

3. Anna Zeide, *Canned: The Rise and Fall of Consumer Confidence in the American Food Industry* (Berkeley: University of California Press, 2018).

4. "Canned Peaches Fatal: Two Boys Die from Ptomaine Poisoning—Mother May Die," *New York Times*, March 18, 1907, 5.

5. Mary Virginia Terhune [Marion Harland] and Christine Terhune Herrick, *The National Cookbook* (New York: Charles Scribner's Sons, 1896), 484–486.

6. Christine Terhune Herrick, "What I Have Learned about Canned Foods," *The Woman's Home Companion*, February 1914, 20.

7. Marion Harland, *The Story of Canning, and Other Recipes* (Washington, DC: National Canners Association, 1910).

8. Herrick, "What I Have Learned about Canned Foods," 20.

9. As cited in Manners Smith, *Marion Harland*, 564.

10. Herrick, "What I Have Learned about Canned Foods," 20.

11. See Nancy Tomes, *The Gospel of Germs: Men, Women, and the Microbe in American Life* (Cambridge, MA: Harvard University Press, 1999). This theme was also prevalent in Harland's *The Story of Canning*, with one image caption reading, "The canner and the scientist work hand in hand in the great work of destroying the germ life" (p. 23).

12. For more on Progressive Era reform, see Michael McGerr, *A Fierce Discontent: The Rise and Fall of the Progressive Movement in America, 1870–1920* (New York: Oxford University Press, 2005).

13. Herrick, "What I Have Learned about Canned Foods," 20.

14. Harland, *The Story of Canning*, 2–6.

15. Zeide, *Canned*, chaps. 1–3.

16. Herrick, "What I Have Learned about Canned Foods," 20.

17. The first two quotations are from the *40th Anniversary National Canners Association* (Washington, DC: National Canners Association, 1947), 10, Grocery Manufacturers Association Library Archive, Washington, DC. "Intensely cold" is from *A History of the Canning Industry* (Baltimore: Canning Trade, 1914), 74.

18. Harland, *The Story of Canning*, 3.

19. *A History of the Canning Industry*, 76.

20. *40th Anniversary National Canners Association*, 14.

21. Megan Elias, *Stir It Up: Home Economics in American Culture* (Philadelphia: University of Pennsylvania Press, 2010).

22. Manners Smith, *Marion Harland*.

23. Marion Harland, *Marion Harland's Autobiography: The Story of a Long Life* (New York: Harper & Brothers, 1910), 344.

24. Richard White, *Railroaded: The Transcontinentals and the Making of Modern America* (New York: W. W. Norton, 2012); Brandon Fox, "Domestic Expert," *Richmond Magazine*, June 17, 2013, http://richmondmagazine.com/restaurants-in-richmond/domestic-expert-06-17-2013/.

25. Jessamyn Neuhaus, *Manly Meals and Mom's Home Cooking: Cookbooks and Gender in Modern America* (Baltimore: Johns Hopkins University Press, 2003).

26. Manners Smith, *Marion Harland*, 537.

27. Marion Harland, *Marion Harland's Complete Cook Book: A Practical and Exhaustive Manual of Cookery and Housekeeping* (Indianapolis, IN: Bobbs-Merrill, 1903), 540.

28. Marion Harland, *Common Sense in the Household,,* (New York: C. Scribner, 1872), 52; Laura Shapiro, *Perfection Salad: Women and Cooking at the Turn of the Century* (New York: Farrar, Straus and Giroux, 1986).

29. Franco-American Food Co., "The Average Cook" (advertisement), *The Judge* 17 (1889): 360; Franco-American Food Co., "In Glass, for the Sick" (advertisement), *Life* 16 (September 1890): 143.

30. Harland, *Marion Harland's Complete Cook Book*, 32.

31. Shapiro, *Perfection Salad*, 201.

32. Harvey Levenstein, *Revolution at the Table: The Transformation of the American Diet* (New York: Oxford University Press, 1988), 156–158.

33. Manners Smith, *Marion Harland*, 494.

34. Manners Smith, *Marion Harland*, 566.

35. In *Just a Housewife: The Rise and Fall of Domesticity in America* (New York: Oxford University Press, 1987), Glenna Matthews also portrayed Harland "as a transitional figure between traditional cook-housekeepers of the mid-nineteenth century . . . and the new home economists of the twentieth century" (according to Manners Smith, *Marion Harland*, 15).

36. Manners Smith, *Marion Harland*, 550.

SCIENCE

·

11

Modern Food as Ranked Food

WHO'S AFRAID OF THE DARK SUGAR?

David Singerman

First a London doctor found the bug.[1] Because it inhabited almost all kinds of sugar, he called it *Acarus sacchari*: the sugar mite (figure 11.1). Under a microscope, each spoonful became a repulsive spectacle of living, dead, and dismembered insects. *Acari* are everywhere, he said, except the most refined sugar; only modern industrial practices could guarantee its absence. The normal processes of manufacture on tropical plantations did not wipe the mite out. Dark raw sugars were barely safe to handle, let alone eat, and in those who traded raw sugar, it caused an ailment called "grocer's itch."

After its discovery in 1855, the sugar mite spawned. The following decade a Scottish pamphlet about the bug found readers among US senators.[2] In 1879 *Popular Science* claimed they swarm 350,000 to the pound.[3]

An ambitious congressman brought the *Popular Science* article to the floor of the House of Representatives. The mite's presence in a nation's sugar, said James Garfield, indexes its progress toward civilization: the cruder the product, the cruder the people. Thus darker-skinned Cubans and Filipinos made sugar in a "black cheap" form full of *Acari*, while white American stomachs could tolerate only refined sugar, free of bugs and other impurities. It took honest American workers, Garfield said, to purge sugar of traces of foreign people like these grotesque invaders.[4]

Americans were eating more sugar each year and needed to protect their bodies from the dark origins of their own food. And the government couldn't guard its citizens from unhealthy sugar, Garfield and others

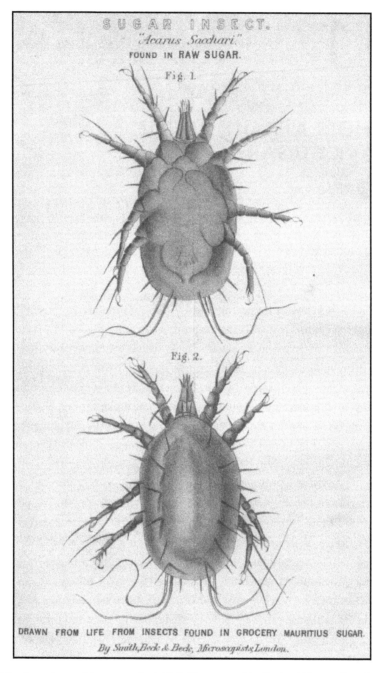

11.1   A specimen of *Acarus sacchari*, 1868. (Courtesy of the National Library of Medicine.)

believed, without also guarding the country's refiners from unfair competition and its finances from unscrupulous smugglers. Dark sugar, as we'll see, had a lot to answer for.

To the US government of the late nineteenth century, almost every foreign object was a threat. Protectionism shaped economic policy, shielding American manufacturers from overseas competition by erecting steep tariff walls. Of course, those manufacturers sometimes needed foreign raw materials to produce finished goods, so while final products from abroad paid high taxes to deter their entry, crude or raw substances entered relatively unscathed. Sugar refining was a giant domestic industry. Companies who sold sugar for American tables made sure that their direct competitors were largely kept out, while the raw sugar their machines needed entered much more cheaply.

But not all finished goods appear equally finished. Protecting a woolen mill was straightforward: it took no special expertise to tell yarn from a jacket. To distinguish between raw sugars and refined ones was harder. So to make sure that raw sugar really was raw, every single sugar cargo had to be directly appraised by an officer's eye and his special instrument after it landed on the wharf but before it was melted in a refinery.

This special instrument wasn't much to look at: just twenty-five glass jars containing the whole spectrum of sugar, numbered in order from a molasses-dark brown (No. 1) to bright white (No. 25). It was called the Dutch scale, because for obscure reasons of the sugar trade it came certified by the government of the Netherlands. Presented with a sample of sugar, an appraiser matched its appearance to one of his jars. Just like the bread and beer in other essays in this book, sugars entering the United States were judged by their color.

More importantly, just like people. Even the spectra of people and sugar were similar: dark to light, scorned to valuable. But unlike people, sugars of lighter color found it harder to cross the border, not easier. To each increment of the Dutch scale, Congress had assigned a level of tariff: one rate below No. 7, a higher rate from 7 to 10, still higher from 10 to 13, and so on. The whiter and more refined the sugar, the higher the cost to bring it into America.

Of course, for the appraiser to make an accurate judgment of a cargo he needed a representative sample of its contents. So, on the sugar docks—America's frontier against dark foreign sugar—stood the sampler, the lowly public servant whose job it was to make sure that sugar didn't sweet-talk its way through the gate.

The customs sampler confronted containers of all sizes and shapes. But rather than blindly stab his sampling tool—called a trier—into the sugar, the sampler first inquired about its journey to the wharf. For example: how had its molasses been drained away? How had the sugar traveled? First-class saccharine passage was the sack, while in steerage most sugar sailed in colossal barrels called hogsheads, jammed into the hold to wring every dollar from a voyage. As soon as sugar went belowdecks, gravity drained it, and any remaining molasses headed straight for the keel. Sugar ships were filthy places, as 10 or even 20 percent of raw sugar's weight seeped out. Dark sugar above could spoil white sugar below, playing tricks upon anyone trying to represent hundreds of tons in just a few ounces.

If the sugar fooled the sampler into thinking it was darker and wetter than it was, the buyer would pay too little in taxes. If the sugar fooled the sampler on the lighter end, then he would be charged too much. It took years for a sampler to learn how to read the "character" of a cargo in order to choose where to strike with his trier. Even then, sampling was more connoisseurship than precision: even after decades on the job, samplers would frequently disagree about the true character of sugar.[5]

The deeper truth was that much like the earthy spectrum of sugar itself, there existed a gradient of plausibility and denial. A sampler could make the sugar say what he wanted. The Treasury was paranoid about bribery by merchants and sometimes regretted that of all the employees in the customs service it paid samplers the least.[6] It would be easy—it was easy—for a sampler to select excessively dark samples for appraisal in exchange for an envelope of cash.[7] Understandably, importers rarely worried about overpaying tariffs.

Even with a superior looking over his shoulder, there was no way to tell whether a sampler was faithfully executing his post. His behavior always looked the same: he asked about the container and its contents, plunged the trier into the sugar, and, with a twist, deposited the tubeful onto a paper. Once he'd captured the character of the sugar, in his practiced estimation, the sampler folded the paper's edges, scribbled the ship's information on its outside, and sent it on its way.[8]

The sugar that arrived in Baltimore aboard the *Mississippi* on the last Monday of November 1877 ought to have been easy to sample and appraise. Its bags came from Demerara, a compact stretch of Guiana coast that commanded disproportionate influence within the sugar world. Demerara's legendary sugar boilers found work across the Caribbean, and its plantations boasted some of the world's best equipment. In recent decades the colony had so eagerly adopted new technologies that it had given its name to one of the most desirable kinds of sugar on the market, a pale-yellow crystal of remarkable purity. These sugars were the property of a Baltimore merchant named William Perot, having been dispatched from South America by his brother and business partner Adolphus.

Perhaps because of the reputation of Demerara sugar, when customs officials unfolded samples from the *Mississippi* they did not expect darkness to stare back. Some of Perot's cargo rated between Nos. 7 and 10 on the Dutch scale—normal for raw sugar, though darker than most from Demerara. But 712 bags were declared below No. 7, making them among the darkest sugars entering America and those paying the least in taxes. But they fit a pattern: for two years customs agents had been noticing sugar from everywhere getting darker. And Treasury officials believed they knew why, even if they could not often prove their belief to legal satisfaction.[9]

Two years earlier, in 1875, Congress had raised rates on sugars of all colors. There were plausible explanations for why importers would gravitate to darker sugar after such a change in the tariff. Higher tariffs might eat into the profit margin of light sugar more than dark, making dark sugar relatively more attractive to merchants. What bothered the Treasury was that even as the average quality of imports had been getting worse, the products of American refineries—made from those worsening imports—stayed the same quality or even improved.

Importers had an explanation ready: a sugar's color no longer bore a direct relation to its character. The manufacture of sugar itself had changed, thanks to the spread of technologies like those first adapted in Demerara. New sugars were not like the old. An antique tariff system animated by ideas about color no longer applied to centrifugal sugars or others made with new machinery. Far more sweetening power could hide under a darker tint these days. But this explanation didn't convince Treasury officials, who noticed, for instance, that buyers and sellers still seemed able to agree on just how much a cargo of sugar was truly worth. If new sugar was really so confusing, surely some commercial buyers— and not just the government—would occasionally question its value? The more likely explanation was that sugar importers were intentionally darkening their sugar to pay less tax.

Such "artificial coloration" had been the conclusion of several Treasury commissions on the subject earlier in 1877, who had warned that the practice was widespread among importers. The commissioners advised punitive tariffs whenever such coloration was discovered. And after years of darkening cargoes, Secretary of the Treasury John Sherman declared that he'd heard enough. The sugar tariff, he said, was obviously being subverted by fraudulent imports, and this crime against the people's finances was an "embarrassment" to his customs service.[10] (Rarely do bureaucracies admit fallibility, but "Sugar, embarrassment in the collection of duties on" got its own entry in the index.) The secretary told his officers to seize sugar that they judged was artificially colored.

Merchants from major ports sent anguished delegations to Washington. Baltimore's emissaries—including the president of one of its refineries, the head of its customs house, and William Perot himself—pleaded with the government that they had specifically instructed their Demerara suppliers not to artificially color their sugar.[11] But the city protested too much. When the New York and Boston commissions warned about artificial colors, the Baltimore equivalent reassured Sherman that the current system was working just fine.[12] The city's federal appointees and its sugar merchants were worryingly friendly. When the Treasury cabled the US consul in Georgetown about artificial coloration, Perot and his fellow Baltimoreans somehow got the consul's reply as quickly as Sherman had. The consul insisted to his superiors that Demerara sugars were innocent, then he boarded one of William Perot's ships and sailed for home.[13]

So the Treasury special agent in Baltimore was probably not surprised when he received information that Perot's bags on the *Mississippi* were artificially colored. A former refiner played tipster, perhaps hoping he would receive a plum job as a reward for intervening in a celebrity case.[14] Speed mattered. By now the bags had been hauled to a warehouse and Perot was busy negotiating with the city's refineries for a sale. As soon as a deal was struck, the evidence—or the perpetrator, depending on your perspective—might be melted. The agent rushed a message to Sherman.

Perot got no warning. Late Friday afternoon, just four days after the *Mississippi* had landed in Baltimore, the Treasury seized 712 of Perot's bags of dark Demerara sugar, 100,000 pounds purporting to be below No. 7 on the Dutch scale.

It had been years since the sugar market bubbled in Baltimore, and the seizure froze it. Those 712 bags represented the majority of the stock in port. Guiana was also a major export market for the city, but Baltimore captains would not sail if they had no hope of returning laden with sugar, and Demerara planters had no interest in shipping sugar they could not sell. When Sherman's men seized another Demerara cargo a month later, this time in New York, the colony more or less stopped its shipments to America.[15]

It was in no one's interest that Perot's case should linger for a year. But the arbitrators named jointly by the merchant and the government failed to agree on the value of the cargo, so the court appointed two commissions: one to travel to Demerara to interview those involved in making it and another to collect testimony from American sugar makers in Louisiana. Meanwhile, Perot reached out to the country's most esteemed chemistry professors for their analyses. But he became paranoid that the government would intercept his sugar samples in the mail, and the chemists tired of his habit—common among Baltimore's merchants in the years after the Civil War—of casting his struggle against federal law as one of justice against tyranny.[16]

After many months, the trial got underway. Perot's attorneys didn't deny that the sugar was meant to be dark. The manufacturers of the cargo readily admitted that they had not tried to lighten it. They had run the juice only through coarse filters, they hadn't skimmed impurities, and they hadn't even followed the common practice of washing the sugar in the centrifuges. These planters and merchants recognized that, broadly speaking, they were taking an odd position. For most of the history of sugar, the money had been in the high end of the market. So makers and sellers had wanted to make sugar lighter in color because, for most of the history of sugar, lighter color signaled higher quality.

But that calculus had changed, they said, when Congress had meddled with the natural order of things. The higher tariff rates meant that lighter sugar was now too costly to import to make any financial sense, given how comparatively cheap it was to make crude dark sugar. In short, they were honest manufacturers who sold innocent sugars, and now the government "seemed to regard as very wicked" that they simply wanted to make sugar that they could sell at a profit.[17] The government, they said, was the party acting wickedly here.

Witnesses for the defense accused Sherman of sending spies to Demerara and of tampering with the samples of sugar that the court-appointed commission had brought back as evidence.[18] Of course, over weeks and months the sugar might just have been "changed by keeping," but it never hurt to insinuate wrongdoing; there was more than one way for a savvy merchant to exploit sugar's instability.[19]

To find Perot guilty, the jury had to affirm not only that the sugars were dark but that they had been colored by artificial means after they were crystallized from the cane juice, and that Perot had intended to defraud the government by bringing them into the country. And the defense refused to concede that the sugar had been colored "artificially." What was the artifice? No extraordinary measures or substances had been used. Indeed, other witnesses testified that the process of manufacture was normal for Demerara—normal, that is, since 1875, when a New York merchant sailed into Georgetown with samples of dark Cuban centrifugals and asked for boilers in the colony who would copy them.[20]

The court, like the samplers and appraisers, found that eliciting the truth from sugar was more difficult than it first appeared. Defense and prosecution witnesses disagreed both about philosophical questions relating to sugar, such as whether a skilled boiler could grow perfect crystals from dirty juices, and on factual ones, like whether a dark-colored matter clung to the particular crystals in Perot's case. Everyone agreed that Perot's sugars lightened when washed, but did this mean that something extra had been added, something both superficial and artificial? This unlucky jury was asked to identify the precise moment at which human ingenuity tampered with nature. Did the washing reveal a scheme to smuggle light sugars into America under the cover of darkness?

Unsurprisingly, the jury's answers satisfied nobody. They found the sugar guilty of being artificially colored. Perot's sugars had experienced something outside a crystal's normal life cycle. But they also rinsed Perot clean, acquitting the importer himself of knowing about any scheme. At the end of the trial he still owned the 712 bags of sugar, and being legally no lighter than No. 7 on the Dutch scale, they owed not a cent more to the Treasury. The dark sugar sorcery of Demerara, the jury decided, had fooled even this experienced merchant.

On the question of artificial coloring Secretary Sherman had been vindicated, but on the matter of enforcement he had been defeated, so he spent the next few years trying to persuade Congress to change the tariff law.[21] To pierce any false color, Sherman wanted to empower customs agents to use sophisticated chemical instruments that, advocates promised, could precisely and accurately quantify the purity of a sample of sugar. And since Caribbean sugar was still being manufactured to subvert the tariff, he still wanted to know how. So he dispatched three Treasury agents back to the Caribbean to discover, once and for all, the distinction between natural and artificial color. Just to remind merchants that he had not forgotten, Sherman made sure the expedition included not only the Baltimore officer who had seized Perot's sugars but also the refiner who had squealed.

They set sail in February 1880, samples of offending sugars in their baggage, on an itinerary that recapitulated the evolution of sugar-making. The first stop was Dominica. Its production, the agents reported, was unsophisticated, from the old-fashioned mills to the "natives" who walked into town carrying the finished sugar on their heads. They investigated Dominican boiling houses and even experimented themselves, seeing what happened if they didn't remove impurities or if they could change sugar's color by washing it after it was finished.[22]

Next was Barbados, the center of the seventeenth century's revolution in sugar production, where the agents learned how planters and boilers adjusted sugar to suit purchasers' tastes. The owner of a plantation admitted that yes, tax-conscious American importers complained when sugar was too lightly colored. Meanwhile, English customers wanted purity but preferred a light yellow color to actual white, so planters added a touch of sulfuric acid to tinge their crystals.[23] One veteran boiler revealed his secret to making high-quality dark sugars: boil the sugar for three days to make perfect crystals and only then add the molasses to coat them. Any earlier and the color could never be washed off the sugar to whiten it for sale.[24]

At last the agents were ready for Demerara, where they learned why the colony's boilers were coveted. Dark sugars made there—dark enough to register below No. 7—turned out to measure 96 percent purity or more under chemical analysis, or as good as the sweetest and brightest centrifugals from Cuba.[25] The true wizardry happened inside the vacuum pan, the sealed boiling chamber where the boilers cultivated crystals and earned their pay. By manipulating the pressure as the juice boiled, they could make molasses "adhere tenaciously" to a crystal's surface, creating sugar far darker than any competitor yet measuring the same under chemical tests. At trial Perot had claimed his sugars were dark because their impurities gave them an inherent tint. These Demerara sugars were carefully produced to take the impurities out and guarantee their otherwise "superior" character.[26]

All along the journey the agents showed samples of Perot's sugars to boilers they met. Fingering just a teaspoon of sugar, these men could decipher the story of its manufacture—not only whether the sugar had been made in a vacuum pan or not, but also their fellow boiler's technique, his method of clarification, how much sulfuric acid he added, and how he liked his molasses. After examining the samples by the eye and under the microscope, these craftsmen reported that there was no chance Perot's sugars were "natural."

By protecting American refiners from light sugars disguised as dark, the tariff also protected white American sugar workers from deceitful foreigners who would steal the rewards of their honest labor. Nearly 6,000 men worked in sugar refineries in the United States, and four times as many in dependent industries: making barrels, working iron, and fixing machinery.[27]

Some peoples represented a threat through the baseness of their products. Take the Philippines, the source of some of the darkest sugar shipped to America. Filipinos, said Congressman Garfield in the same speech where he brought up *Acarus sacchari*, lacked the intelligence to make sugar that might compete with American refined products. They boiled sugar in the "crudest, rudest, simplest way, by labor the cheapest and least skillful," and the sugar they made was a "black cheap form," "the dirtiest yet known." Only in America, by American craft, could guileless sugar from simple Filipinos become free from invisible bugs and fit for the civilized table.[28] On the one hand, foreigners were mocked for their supposed stupidity: it took no ingenuity to make raw sugar the old-fashioned way. "The foreign labor of gravitation on the soil of Cuba" drained the molasses out of the sugar, said an Illinois congressman. "That is all the labor there is in it."[29] The parallel between how the Dutch organized their sugar scale—dark to light, lowest to highest—and how the West constructed its worldwide racial hierarchy was obvious enough that it became an official joke. Garfield asked his honorable friends to imagine a hogshead sitting still for a month before a sampler bored into the top and the bottom. The samples would be lighter and drier near the top, wetter and darker below. "How much foreign labor," the antislavery advocate Benjamin Butler inquired, "is there in the settlings of that sugar?" "There certainly is a good deal of dirt in it," Garfield replied, and he and Butler shared a laugh.[30]

On the other hand, foreigners who were too clever served as a justification for ditching the Dutch scale in favor of chemical tests. In a congressional hearing room a former Treasury agent unwrapped two samples of imported sugar, one dark and one white—no less than 20 on the Dutch scale. He then pulverized the former into the latter, right in front of the committee, to show how clever makers could disguise light sugar's true nature.[31] Without new means of detecting the true sweetness of imported sugar, America would only bring in "the sugar that has been advanced by the higher and more intelligent processes of our nearer neighbors."[32]

Imaginations raced about what else foreign sugar might contain. Conditions in Cuba, Americans read in the papers, were so miserable that enslaved people chose suicide by sugar cauldron. And Chinese indentured workers who died in other ways might find their skeletons exhumed and burned, their char mixed with that of livestock to better filter the juice. The sugar mite and the grocer's itch were minor irritants compared to the horror of sugar adulterated with the corpses of coolies.[33]

But sympathetic or not, foreign workers were still dangerous. Pamphleteers for US refiners prophesied huge new refineries constructed in

Havana if the tariff laws were not reformed.[34] Many of America's sugar
workers had, not long ago, fought and bled to defeat the slave power.
They were certainly not about to give up their jobs to enrich Cuban slave-
holders now. And there were better ways to liberate Cuban slaves than to
import the island as impurities in hogsheads, conquering each pound of
the colony's dirt at the grocery price of sugar.[35]

For all the twenty-first-century efforts to reassert local food networks in
the United States, our demand for stimulants like sugar, tea, and coffee
will always be sated by distant suppliers. The anxieties over dark sugar in
the 1870s show how that dependence means that people, them and us,
are as vulnerable to manipulation as substances themselves. We want to
be conscious of the unseen origins of our faraway food, but in truth we
know little about where it comes from.

The fact that sugar is a ubiquitous food made it especially easy to manufacture those fears. Consider the sugar mite. Had rumors spread of a tiny
bug infesting another commodity—bituminous coal, say, or lumber—the
reaction among consumers would have been far less visceral. It's hard to
imagine Garfield speechifying that the presence of some invisible insect
in its ore supply attested to a nation's backwardness. The substances we
choose to ingest carry more meaning, for us and for others, than substances we keep external. We speak of "consuming" all sorts of products,
but we only really consume food. Food is different from other goods.
That's why you picked up this book.

The tariff protected American refiners, and refiners wanted that tariff
enforced, so they encouraged white Americans to distrust dark tropical
sugar. Once they had digested that first lesson they graduated to the second: that dark sugar freely exchanged its troubling characteristics with
the human beings who made it. Sugar's ubiquity as a food meant that a
racist coating was easily absorbed into the blood. But though food may
be especially ready to exchange its character with people, it's not alone.
Money can and does buy the same sort of story denigrating other goods,
other places, and other people. We're told that electronics from China
imitate our technology and spy on us, just like Chinese firms and people.
Cheap junk comes from Donald Trump's "shithole" countries.

Whenever we eat food from over the horizon, we rely on others to tell
us what to think of it—whether that food has the pastoral resonance of

sugarcane or, as Ben Cohen tells us in his essay (chapter 6), the industrial twang of glucose. And those whom we trust may not repay that trust. Sugar chemists were laughing about *Acarus sacchari* decades later. The sugar mite, with its nasty proboscis and hooked appendages, turned out not even to exist, some hybrid of convenient error and intentional hoax. Still, good luck unseeing it. Enjoy your next packet of Sugar in the Raw.

There is a curious coda to the American government's pursuit of Perot's dark sugar, another tale of artificial coloration and deceptive arrival. The story would be a parallel one, but for the fact that it actually intercepts the *Mississippi* and her cargo.

By 1878 America had been gripped by the search for Charley Ross for four years, ever since the toddler was snatched from the yard of his Philadelphia home. The suspected kidnappers died in a shootout but left few clues, and would-be Charleys materialized around the country. The most famous pseudo-Charley turned up that January, in Baltimore of all places, and aboard a ship from Demerara. Charley's father checked off the five hundred and seventy-third sighting of his son. As one headline put it, "Charley Ross—Another of Him Found."[36]

The mystery was how Charley #573 had wound up in Demerara at all, let alone living, as newspapers reported, with a disreputable mulatto woman. She claimed that she was born in Demerara, moved to the United States, and married a white man, who promptly disappeared. She sought work in Boston, where rumors placed her with a known accomplice of the kidnappers, before returning to South America with the child. Some reports suggested she abused him, and that from the moment they arrived in Demerara the boy insisted he'd been stolen. And then Adolphus Perot recognized the child on the street, fed and clothed him, and "consigned" him to his brother William, just like cargo (figure 11.2).[37]

This boy was "the most probable Charley Ross yet," said one paper, and a crowd gathered at the docks just to watch him alight to buy a hat.[38] But this cargo's papers proved suspect. The boy spoke crude "foreign" English and called himself George Sylvano Signio Rio, although he supposed he might once have been called Charley Ross. His skin and hair were also darker than Charley's. After reading that the boy has "an olive complexion peculiar to those who live in a warm climate," the exhausted Ross family rejected the possibility that he was their son.[39]

11.2   The face Adolphus Perot saw in Demerara. (Courtesy Lester S. Levy Sheet Music Collection, The Sheridan Libraries, Johns Hopkins University.)

Adolphus Perot had also forwarded some money for his care, so the upbringing and education of this false Charley were entrusted to the wonderfully named Home of the Friendless. Under American care—one might say refinement—at this wholesome institution, the dark foreign boy underwent a miraculous change in his outward presentation. His accent evaporated as he began to recall details of early life in Philadelphia. He revealed that he had been artificially colored in Guiana, where he was forced to bathe apart from other children "in a peculiar kind of dirty water." His identifying marks—from vaccinations—became visible. "The swarthiness of his complexion was wearing off," a correspondent exulted, "and he was becoming whiter."

Now the Ross parents liked what they saw in a photograph. Their doctor visited Baltimore and summoned the family, for the child was "growing to look more and more like Charley." When the father arrived, the Home for the Friendless kept him away until Charley's appearance "improved." They wanted "his natural color somewhat restored, as his skin was assuming a fairer color, the dark hue gradually disappearing." The child in their care, they felt, ought to make the sweetest possible impression on his appraiser.[40]

After interrogating the boy for a few hours, however, the father left without a word to the press.[41] Under all the coating, the boy was just another import from Demerara. He wasn't to be trusted, wasn't fit for white tables. Perot's young charge, like his bags of sugar, had been disguised in the tropics to slip into America.

NOTES

1. Arthur Hill Hassall, *Food and Its Adulterations: Comprising the Reports of the Analytical Sanitary Commission of "The Lancet"* (London: Longman, Brown, Green, and Longmans, 1855), 17–19.

2. Robert Niccol, *The Sugar Insect: "Acarus sacchari," Found in Raw Sugar* (Philadelphia, 1868). The HathiTrust copy from Harvard University Library indicates that it was gifted to the library in 1869 by Charles Sumner.

3. E. R. Leland, "Mites, Ticks, and Other Acari," *Popular Science Monthly*, February 1879, 508–509.

4. April Merleaux, *Sugar and Civilization: American Empire and the Cultural Politics of Sweetness* (Chapel Hill: University of North Carolina Press, 2015).

5. "The Custom-House Inquiry: Further Evidence by Examiners," *New York Times*, May 23, 1877.

6. *Annual Report of the Secretary of the Treasury on the State of the Finances for the Year 1878* (Washington, DC: Government Printing Office, 1878), xxviii.

7. *Testimony in Relation to the Sugar Frauds, Taken by the Subcommittee of the Committee of Ways and Means of the House of Representatives, New York, September 1878* (Brooklyn, NY: Brooklyn Daily Times, 1878), 36.

8. *Testimony in Relation to the Sugar Frauds*, 8.

9. *Annual Report of the Secretary of the Treasury on the State of the Finances for the Year 1877* (Washington, DC: Government Printing Office, 1877), xxvi; F. W. Taussig, *The Tariff History of the United States, Part I*, 5th ed. (New York: G. P. Putnam's Sons, 1910), 142.

10. *Annual Report of the Secretary of the Treasury on the State of the Finances for the Year 1877*, xxvi.

11. "Colored Imported Sugars: A Seizure by the Government," *New York Times*, January 26, 1878.

12. "Commissions to Examine Certain Custom-Houses of the United States," US House of Representatives executive document no. 8, 45th Congress, 1st session (1877), 1–10.

13. "The Coloured-Sugar Question in America," *The Sugar Cane*, February 1, 1878, 64.

14. *A Review of the Case of the United States vs. 712 Bags of Dark Demerara Centrifugal Sugars: A. W. Perot & Co., Claimants: Tried at the September Term of the District Court of the U. S. for the District of Maryland* (Baltimore: Lucas Brothers, 1878), 4, 14; Nicholas Parrillo, *Against the Profit Motive: The Salary Revolution in American Government, 1780–1940* (New Haven, CT: Yale University Press, 2013), chap. 6.

15. "The Coloured-Sugar Question in America," 66; "Colored Imported Sugars."

16. W. H. Perot to Charles Chandler, October 1, 1878, Charles F. Chandler papers, Box 260, Folder 7, Rare Book and Manuscript Library, Columbia University Library (hereafter Chandler papers). For correspondence between Perot, Chandler, and Morton, see Chandler papers.

17. *A Review of the Case of the United States vs. 712 Bags of Dark Demerara Centrifugal Sugars*, 30.

18. *A Review of the Case of the United States vs. 712 Bags of Dark Demerara Centrifugal Sugars*, 53.

19. W. H. Perot to Charles Chandler, April 11, 1878, Chandler papers, Box 260, Folder 7.

20. *A Review of the Case of the United States vs. 712 Bags of Dark Demerara Centrifugal Sugars*, 28.

21. *Annual Report of the Secretary of the Treasury on the State of the Finances for the Year 1878*, 29.

22. S. E. Chamberlin et al., *Report on the Methods of Manufacturing Sugar in the West India Islands and British Guiana* (Washington, DC: Government Printing Office, 1880), 3–4.

23. Chamberlin et al., *Report on the Methods of Manufacturing Sugar*, 12–16.

24. Chamberlin et al., *Report on the Methods of Manufacturing Sugar*, 38.

25. Chamberlin et al., *Report on the Methods of Manufacturing Sugar*, 16, 21.

26. Chamberlin et al., *Report on the Methods of Manufacturing Sugar*, 23.

27. *1880 Census*, vol. 2, *Report on the Manufactures of the United States* (Washington, DC: Government Printing Office, 1883), 77, https://www2.census.gov/library/publications /decennial/1880/vol-02-manufactures/1880_v2-01.pdf; James A. Garfield, "Sugar Tariff: Speech of Hon. James A. Garfield, of Ohio, delivered in the House of Representatives, Wednesday, February 26, 1879" (Washington, DC, 1879), 5–6.

28. Garfield, "Sugar Tariff," 12.

29. *Review of the Efforts of the Forty-Fifth Congress of the United States for the Revision of the Sugar Tariff, together with Speeches Made by Members of the House of Representatives* (New York, 1879), 9.

30. Garfield, "Sugar Tariff," 8.

31. Henry Brown, *Statements Made before the Committee of Ways and Means in the Interests of American Consumers, Home Industries, and Revenue* (Washington, DC: Judd & Detweiler, 1880), 22–24.

32. Garfield, "Sugar Tariff," 12.

33. "Suppression of the Coolie Trade," *New York Times*, February 26, 1877.

34. David Ames Wells, *The Sugar Industry of the United States, and the Tariff: Report on the Assessment and Collection of Duties on Imported Sugar, On the Results of an Economic and Financial Inquiry into the Relation of the Sugar Industry of the United States in Its Several Departments of Production, Importation, Refining and Distribution of Product, to the Existing Federal Tariff* (New York, 1878).

35. *Testimony in Relation to the Sugar Frauds*, 99.

36. Bill James, *Popular Crime: Reflections on the Celebration of Violence* (New York: Scribner, 2011), 39; "Charley Ross—Another of Him Found," *New Orleans Daily Democrat*, January 20, 1878.

37. "Charley Ross—Another of Him Found"; "Charley Ross," *Alexandria Gazette and Virginia Advertiser*, February 5, 1878; "Charley Ross," *New York Herald*, February 5, 1878.

38. "Charley Ross—Another of Him Found."

39. "Is the Boy Charley Ross?," *New York Times*, February 3, 1878.

40. "Charley Ross," *New York Herald*, February 5, 1878.

41. "Not Charley Ross," *New York Herald*, February 6, 1878.

12

Modern Food as Extracted Nutrition

DARBY'S FLUID MEAT, DIGESTION, AND THE BRITISH IMPERIAL FOOD SUPPLY

Lisa Haushofer

Frederick William Pavy had a beef with beef extracts.

This was most inconvenient because he was surrounded by them.

In the nineteenth century, the pages of medical and pharmaceutical magazines abounded with advertisements for products with such descriptive names as Justus Liebig's Meat Extract, Gail Borden's Meat Biscuit, and Gillon's Essence of Beef. Recipes for preparations of beef extracts, beef teas, and beef broths adorned the food sections of nursing manuals, cookbooks, and popular magazines. Specimens of various concentrated meats were exhibited in the growing number of mid-nineteenth-century public display spaces, including London's Great Exhibition of 1851, the Paris Universal Exhibition of 1855, and the Museum of the British Pharmaceutical Society.

But to Pavy, as a physiologist, they were all wrong. "In beef tea, broths and the extracts of meat prepared in the ordinary way," he lamented, "we have physiologically a very imperfect representation of an article of nourishment."[1] He thought there had to be a better way to reduce food to its most nourishing components, a way that could only be found through his scientific field of physiology, in contrast to the field of chemistry that had given birth to meat extracts.

Enter Darby's Fluid Meat, a new product that was the first "artificially digested" food, born of a process that mimicked the physiological process of digestion in the human body—or so Pavy claimed.

Despite Pavy's strong disapproval of one—the standard extracts—and support of the other—the new fluid meat—these two meat products were both concentrated forms of food, something like our modern-day bouillon or stock cubes.

But in their time these products represented something much more. Especially in the British context, at a time of imperial conquest and widespread concerns about having enough food to feed a growing population, using the scientific disciplines to figure out the building blocks of food became all important. Which method of concentration—and which attendant field, chemistry or physiology—would most contribute to the British imperial enterprise became a subject of much debate in the nineteenth century. This debate, at its heart, was about how to distill the nourishing *essence* of food and, in the process, reinvent food as an assembly of functional and not-so-functional (and therefore expendable) component parts. Doing so, the debaters believed, could serve crucial economic and political ends.

Pavy the physiologist's condemnation of meat extracts was partly an act of professional grandstanding. Meat extracts, to Pavy's chagrin, were largely the domain of chemists.[2] Chemists had long been attempting to find the nourishing essence of food, to distill food down to its most functional core. Their efforts traced back to a time when military might was the determining tool of states' power, and when creating foods that could travel along with soldiers, sailors, and surveyors was a central aspect of military and imperial success—like the canned food later described in this volume by Anna Zeide (chapter 10), which was first created in response to Napoleon's need to feed his troops.[3] Chemists had used their expertise to attempt to make foods durable and transportable. In doing so they had come across the problem of reducing foods' mass and weight while still retaining all of its nourishment. They attempted to solve this problem by using chemical extraction methods to produce all sorts of concentrated foods, or food extracts.

Underneath chemists' efforts to produce nourishing extracts of food lay a rather momentous assumption: that chemical methods could imitate processes of life. Specifically, by subjecting foods to chemical extraction and claiming that they could produce the nourishing essence of foods, chemists indirectly suggested that the process of digestion was basically a chemical extraction process. After all, both isolated the nourishing essence of food.

Pavy, along with many physiologists, objected to this way of thinking. He believed in a fundamental difference between life processes and those performed in the laboratory. And so he argued that digestion and the essence of nourishment had to be investigated through physiological methods, those that respected the exceptional nature of life processes.

During Pavy's lifetime the very question of the nature of nourishment achieved a profound political and economic significance that escaped the immediate professional concerns of different groups of scientists. Colonization and industrialization—the transformation of land to concentrate property and maximize manufacture—had made the supply of food into an urgent problem for Pavy's native Britain, a problem that included new dimensions of distance and new scales of need, but also new opportunities of supply.

Whereas food extracts as ship provisioning and army rations had been about making food more portable for travel, nineteenth-century food products were increasingly created as a means of bringing food *closer*. The goal was not so much military conquest elsewhere but an abundant supply of nourishment at home. Food extracts were thus part of a profound shift in imperial strategy, one that was to have lasting consequences for the ecological and economic landscape of Britain's colonies. They were the material manifestation of an increasingly targeted and sophisticated policy of imperial extraction—taking only what served the British Empire and its people and leaving the rest behind.

This is where William Pavy's "fluid meat" comes in. This unappetizingly named item was a new approach to the food problem. The production of "Fluid Meat" allowed Pavy (and others) to frame digestion as an economic variable in considerations of imperial food procurement. The story of the fluid meat, then, much like the story of modern food itself, was not so much about seemingly innocuous technological advances or superior scientific insights. Instead, it was about the powerful logics of exploitation and economization that undergirded colonial and industrial food projects and shaped modern food regimes.[4] Physiologists, too, and not just chemists, could be agents of empire and servants of the British economy, and thus intervene in the burning food question of the mid-nineteenth century—how to secure an abundant food supply.

Looking back at the generations before Pavy can help put his work into the context of those longer efforts to secure food. Since the late eighteenth century, Britons had lived in heightened fear of food scarcity. Common lands had been enclosed, mostly for capitalist manufacture. This decreased the overall domestic production of food, which, in turn, exacerbated unequal access to nourishment.[5] A series of wars at the beginning of the nineteenth century not only brought extended periods of food shortages, but also legislation to protect the interests of domestic grain producers. As David Fouser explores in chapter 2, these measures—the Corn Laws—kept prices high and thereby contributed to growing food inequality. Together with a number of bad harvests in Britain between 1839 and 1842, and a significant economic depression beginning in 1839, this led to some serious national soul-searching on the question of the food supply.[6]

All the while, Britons still lived under the shadow of the dire predictions of the Reverend Thomas Malthus who, in 1798, had articulated the frightening prospect that the population of a nation increased much more rapidly than its food supply. This prompted some drastic measures intended to keep population growth in check, most famously the abolishment of parish relief for the poor (giving poor people "welfare," some argued, would only interfere with the mechanisms that "nature" had designed to keep the population proportionate to its food supply, such as death by starvation).[7] Malthusian reasoning also profoundly shaped British attitudes toward the Great Famine in Ireland in 1845, which caused unprecedented levels of starvation and emigration. The almost tabloid-like newspaper coverage of the famine, while it may not have elicited sufficient concern about the situation in Ireland, kept the dangers of an unstable food supply fresh on British breakfast tables.[8]

Food, in other words, was not just a logistical problem but a pressing political and social issue by the 1840s and 1850s. The French Revolution, in the minds of many nineteenth-century political and economic thinkers, had revealed the dire consequences of an insufficiently fed population, and Britain was witness to its own food riots in the late eighteenth and early nineteenth centuries.[9] Food scarcity was also linked to disease, and therefore to military defeat, as the high mortality from disease among British soldiers during the Crimean War in 1853 seemed to attest. As a result, British thinkers, scientists, and agents of the growing British state tackled the food question with unprecedented urgency and ambition.

Frederick William Pavy was one of them. If you encountered him on the street, you might well have been intimidated by the precision of his gaze, the imposing volume of his sideburns, and the thinness of his mouth, which, even when he was in good spirits, always seemed to display a degree of disapproval. If you looked below his face, however, you would detect the asymmetry and age of his clothes, which betrayed his humble upbringings. The son of a French maltster who had emigrated to England and married the daughter of a small landholder, Pavy was the brother of one older and seven younger siblings. His father died after the birth of the last child and left the family without a regular income. By then the young Frederick William was already firmly on the path to an income of his own, not least through his own strong will and ambition, which were evident to his contemporaries from an early age.

After Pavy twice ran away from his local school as a child, his parents sent him farther away, to London's Merchant Taylors School. There pupils were advanced according to their ability, and Pavy progressed with remarkable speed, traversing the entirety of primary education in only three years. After a time of uncertainty and drifting, in 1847 he finally settled on medicine as a profession and entered London's Guy's Hospital and the University of London.[10]

Pavy rose quickly within the increasingly permeable professional structures of nineteenth-century British medicine. Three driving forces came together for his personal and professional growth. One was a set of new institutions cropping up, including University College London, which operated by offering instruction to those who were not part of the Anglican elite (unlike the traditional universities Oxford and Cambridge). The growing field of public health was another development that helped Pavy.[11] Physiology was the third driving force in launching Pavy's career. In this expanding and increasingly well-situated field, he was particularly interested in food and the physiology of digestion and wrote two lengthy books on the subject.[12] From his vantage point as a physiologist, Pavy felt entitled to make authoritative claims about pressing matters of the British state, including the food supply. He decided to give the chemists a run for their money.

The backstory here is critical. As described above, meat extracts and chemists were intertwined with the British Empire. Preservable, portable meat provisions had a long history of enabling wars, territorial exploration, and imperial expansion. But during the eighteenth and nineteenth centuries, meat extracts were increasingly imagined as important means of guaranteeing a stable food supply *to* a nation at home, rather than just to a nation's emissaries on the move. As economic thinkers argued that a nation's chief means of survival was its ability to generate wealth, national landscapes were transformed to create private property and to prioritize the production of lucrative export goods, such as cotton or wool. Foods became commodities, produced for distant markets rather than for domestic use. In turn, the supply of locally available food decreased. Rural dwellers were driven to cities to find work, further exacerbating the logistical and material challenges of food availability and distribution. Food, it seemed, was increasingly out of reach.

Today we struggle once again with the question of the food supply as climate change, global capitalism, and pandemics have exacerbated food inequality, creating malnutrition, economic instability, and social unrest. To Britain's Victorian technocrats, however, the solution seemed obvious: Britain had to *extend its reach* to find food resources, forage further afield, make use of previously unused edible material, and make a science of knowing which parts of foods provided the most nourishment. Luckily Britain had an empire at its disposal, and British chemists and naturalists played a role in reimagining the imperial map as a "treasure trove of untapped wealth and resources."[13] They surveyed imperial territories for what they portrayed as new and underutilized sources of food, devised schemes for planting particularly nourishing crops, and conducted elaborate calculations that compared foods according to their weight, nourishing potential, and cost.[14]

Meat was considered a particularly rich form of nourishment. Beef had long been the cultural core of the British diet, and nineteenth-century scientific investigations into the nature of nourishment seemed to confirm its primary importance. Meat was also in large supply in places like Texas, Argentina, and New Zealand, which British naturalists and chemists seamlessly included in their imaginary imperial maps, despite the varied ties through which those places were connected to the empire.[15] There was just one problem: how could a substance as perishable and as bulky as meat be transported efficiently and cheaply to British shores from such faraway places, and in sufficient quantities to make the transport cost efficient?

One answer was to transport only those parts of foods that were truly nourishing, thereby reducing the weight, mass, and cost of transportation of food. The search for new and better sources of food therefore comprised, and gave new life to, efforts to identify the nourishing essence of food and separate it from foods' supposedly non-nourishing "bulk." But *what* exactly the nourishing essence of food was, and *how* one could obtain it, was a matter of intense debate.

Chemists thought they had a particularly strong claim to authority over nourishment. They were manipulators of matter, transforming substances by trade. As the food question became ever more pressing, chemists used the method of solvent extraction to isolate what they believed were the nourishing components of all foods, including meat. New products like meat extracts were a crucial part of this investigative enterprise, with Justus von Liebig's meat extract perhaps the most notable example.

Liebig was the century's most famous chemist. That's saying a lot in a century when chemists had become so important to political and economic agendas that you could rank their fame. He created his extract amid raging debates over preparation: specifically, how did boiling, roasting, or salting affect the nourishing constituents of meat? If meat was to nourish, Liebig suggested, somewhat obviously, it had to be prepared so that none of its nourishing constituents would be lost.[16] But alas, salting meat withdrew many of the nourishing components, whereas boiling meat transferred most of the nourishing substances to the boiling water. So Liebig ventured to find a more efficient method of "extracting" the nourishment from meat. He argued that a combination of soaking the meat in cold water (essentially a solvent extraction), slowly bringing the mixture to simmer, and filtering and evaporating the whole to a "dark brown, soft mass" would produce an extract that contained, in highly concentrated form, the nourishing components of meat (thirty-two pounds of ox made one pound of extract!).[17]

Chemically extracted concentrates of meat were therefore a powerful statement by the chemical profession that the nourishing parts of foods *could* be reliably obtained through chemical extraction methods. This was a bold claim as it had direct implications for how the process of digestion itself was imagined. Liebig insisted that since animals could not produce the substances they were composed of, *and* since they received nourishment from the blood, *and* since the blood was composed of substances obtained from food, those substances *had* to be contained, already perfectly formed, in the food animals ate.[18] If precisely those preformed nourishing constituents of food could be obtained outside the body through chemical extraction, the process of digestion was essentially a chemical extraction process.[19] It was a mere separation of preformed, indivisible nutrients from non-nutritious matter.

The physiologists disagreed. They believed that digestion involved a true transformation of matter rather than being merely a process of chemical extraction. That transformation was unique to the body of animals and could not be replicated outside of the physiological surroundings in which life processes took place. Investigations of how the body obtained the nourishing components of food, and what those nourishing components were, were therefore best conducted using physiological methods.

Physiologists insisted that chemical extraction was an inadequate representation of the digestive process. They did this by reference to a recent description by physiologists of a group of substances that acted as transformative agents: the so-called *digestive ferments*, or what we refer to today as *digestive enzymes*. Digestive ferments had first been described by German physiologist Theodor Schwann. He believed in the existence of a digestive "principle," a material substance of the stomach, likely residing in its walls, that stimulated the changes of digestion. In 1835 he set out to identify it. He took the third and fourth stomach of an ox, cut it in small pieces, added water and hydrochloric acid, and waited twenty-four hours. As one does. Then he filtered his mixture, producing a murky yellowish liquid. This he added to pieces of egg white and observed their transformation. Satisfied that he had captured the transforming agent, he proceeded to characterize it by demonstrating its physiological behavior against a set of reagents.[20]

Rather than invent a new name for this type of substance, Schwann decided to "extend the concept of fermentation" and named the agents he had described "digestive ferments." Since the specific substance he had characterized in his experiments "truly effects the digestion of the most important animal foods," Schwann gave it the name *pepsin* (from the Greek πέπτειν or *peptein*, "to digest").[21]

Digestive ferments allowed physiologists to witness the process of digestion in a new way. Its rapidity could be timed and its transformative capacity measured. Physiologists set out to assess the speed of change occasioned in the presence of digestive ferments and determine its relationship to the amount and kind of substance acted upon, as well as the quantity of digestive ferment used. They examined the nature and composition of foods prior to and after having been exposed to digestive ferments, and marveled at the totality of transformation: the substances produced through digestive ferment action, they concluded, were substantially different from the original matter subjected to ferments; they were not merely the disintegrated constituents of a food subjected to chemical extraction.[22]

Pavy's emphasis on the *physiologically* inadequate nature of beef tea, broths, and conventional extracts of meat echoed this conviction in the powerful transformative capacity of digestive ferments and, by extension, of the digestive process. We might not recognize it now, but these were fighting words. They were a thinly veiled stab at the inadequacy of chemistry to examine and reproduce life processes. But the digestive ferments also gave Pavy the idea that the extraction of the nourishing essence from food might be reproduced physiologically by simply imitating the process of nourishment extraction that happens inside the human body through the power of digestive ferments.

And so, in the spring of 1867, Pavy approached a chemist and practical pharmacist named Stephen Darby. At first their conversation centered on the possibility of creating special foods for diabetics. But Pavy then casually mentioned that he had long been struck by "how effectually meat can be dissolved by the process of artificial digestion." He had long wondered, he told Darby, "that it has never been turned to practical account for the purpose." He was convinced that "some day or other the time must arrive for it to be so."[23]

Darby took the hint. In the back of his Leadenhall Street pharmacy he finely sliced a chunk of lean meat into small pieces. He added water, a splash of hydrochloric acid (an acid produced in the stomach) and pepsin. At 96–100° Fahrenheit he let the mixture "digest" until the meat had dissolved. He then filtered the liquid through parchment paper, getting rid of unnecessary "fat, cartilage, or other insoluble matter."[24] Finally, he evaporated the result of his efforts until it acquired the appearance and consistency of "a light brown very thick treacle."[25] With "a strong salt taste" and "an agreeable meaty flavor," the preparation was ready to be ingested; it formed, in Darby's opinion, an "acceptable and very agreeable article of diet."[26] In a pamphlet published in 1870 he proudly announced that he had created "a new preparation of meat."[27] Pavy described the product and Darby's efforts in creating it in detail in his 1867 treatise on digestion.[28]

Darby's Fluid Meat, the first "artificially digested" food product, was born. The process that Pavy had suggested to Darby, and that Darby had followed in preparing his fluid meat, was that of essentially imitating the process of transformation performed by the digestive ferments inside the human body. The "digestion" of the meat with pepsin and hydrochloric acid at a certain temperature produced changes, according to Darby, that coincided "precisely with those which physiologists tell us occur in the stomach in normal digestion when the food has been acted on."[29] His announcement that he had secured a new *preparation* of meat was therefore not only a claim to a new product, but also to a new, *physiological* method of preparation.

Pavy, Darby, and their fluid meat typified a new way to understand digestion. Creating physiological extracts of meat with the help of digestive ferments did a few things. One was that it gave space to physiologists like Pavy, with new preparations like Darby's fluid meat, to effectively critique chemists' methodology. Another was that physiologists began intervening in the chemical approach to making foods available to British tables in the least bulky, most nourishing, concentrated form. Because their intervention came with a different view of digestion, it also had broader implications for the imperial-economic project of optimizing the procurement of food. The introduction of digestion as an important variable in calculations of nourishment allowed physiologists to critique chemical notions of digestion as mere extraction and to attack chemists' equivalence of the constituents of food with the constituents of the body as too simplistic.

Where chemists had claimed to have identified the nourishing essence of food, physiologists cautioned that chemical constituents and physiological nutrients—constituents that could actually be used by human bodies—were not identical. "Although the proximate principles of the blood are chemically the same as the proximate principles of our food," Pavy cautioned, "yet physiologically there is an essential difference between them."[30] That distinction mattered—to Pavy, his colleagues, his antagonists, and to the Empire.

Chemists had drawn up tables listing the quantity of foods' nourishing constituents in relation to their cost. They promoted their efforts as valuable means to plant, procure, and produce more efficiently. Physiologists, in contrast, ranked foods according to their *digestibility* inside human bodies. They scorned chemists for not having taken the variable of digestibility into account. Chemists' calculations of nutrient content in relation to cost of transportation were all well and good, said the physiologists, but they did not actually indicate which parts of foods would be taken up by individual bodies and which parts would be cast out unused, and end up as excrement in Britain's newly built sewers.

Public health officials joined the physiologists in their critique. Edward Smith (1819–1874), a medical officer of the Poor Law Board, a poor law inspector, and himself a physiologist, argued that political economists of food had to take into account "what proportion of a given food passes off by the bowel unused, and therefore what proportion is applied to the nourishment of the body, and what is cast out as useless."[31] The cost-to-nutritiousness ratio of foods on the imperial map, in other words, had to be complemented by the nourishment-to-shit ratio of foods inside bodies.

Emphasizing digestibility meant that, over time, scientists and politicians changed how they approached the food question. Whereas midcentury efforts to increase the food supply had focused on the economy of food procurement and production, attention shifted in the 1870s and 1880s to food consumption as a process in individual bodies. Eaters were instructed in the selection, preparation, and economy of food. An entire machinery of pamphlets, exhibitions, lectures, and institutions taught individuals how to function as economizing extensions of a global resource management program—a true *homo economicus*, or an agent of *domestic economy*. Trifling as they might seem to some, such measures, in the minds of reformers, could potentially determine the nation's economic and political fate. The "Skeleton in the National Cupboard," warned one sanitary scientist, was no longer an "actual deficiency of supply," but the inability "on the part of consumers to lay by stores of supply."[32] What was at stake was nothing less than a shift from a production-centered food economy to one primarily focused on consumer behavior and individual responsibility.

This thinking extended to the very interior of the bodies of individuals. Faulty digestions were a drain on the national food economy as they allowed valuable nourishment to go to waste. But with the new confounding variable of digestibility also came a new plane for intervention. In this context, artificial digestion provided not only a new model for understanding human digestion but also a means of optimizing it. Darby and Pavy had launched a trend: theirs was the first in a series of commercial digestive-ferment preparations intended to not only produce the nourishing essence of foods more physiologically (and therefore more accurately) but to enhance defective digestive systems in individual human bodies.

Following on the new product of Darby's Fluid Meat, a veritable flood of digestive ferment-based products and so-called artificially digested foods filled the pages of popular and specialist publications and the shelves of British pharmacies and chemists in the last decades of the nineteenth century, much as the various meat extracts had done previously. The Mottershead Company marketed Liquor Pepticus, a preparation of pepsin; Liquor Pancreaticus, containing the digestive ferments of the pancreas; and Benger's Self-Digestive Food, a wheat-based food to which pancreatic ferments had been added. Savory and Moore supplied several preparations of digestive ferments, including pepsine and pancreatine. Visitors to the 1884 International Health Exhibition could also admire the company's "Peptonising Apparatus," by means of which "artificial digestion may be readily carried out in the home."[33] "Peptonizing" was a synonym for artificially digesting, peptone being considered the end product of protein digestion. Across the pond, one particular such pepsin-containing recipe would become a stalwart of the twentieth century shelf in the form of Pepsi-Cola.

Digestive ferment products turned the political economy of foods into a practice that could be performed by individuals. The emphasis on digestibility they embodied placed individual digestions at the center of imperial food procurement strategies and shifted responsibility for the food supply from governments to eaters (or rather, digesters). But this was no longer the idiosyncratic digestibility of the highly individualistic body that required the specialized consideration of personal physicians, as had previously been the case. Instead, the digestibility of the nineteenth century was a population-wide variable that had to be taken into account in economic calculations of the food supply. It could thus be optimized through uniform interventions executed by the economic subjects who were doing the eating and digesting.

At once a laboratory experiment and an article of consumption, Darby's Fluid Meat embodied the taste of the time to solve complex social problems with laboratory, often commercialized, solutions. The product and the laboratory method that created this new product were at the center of hard-fought debates about scientific methods to investigate life processes and technocratic approaches to global resource scarcity. Fluid Meat was an argument in those debates: it proposed that physiology was superior to chemistry in solving the food crisis. It also changed the terms of the debate by reframing the problem as one of individual responsibility. In the process, digestion was reinvented as a variable in economic calculations of the food supply.

We no longer feel ourselves connected in the same way to the political economy of food through our digestive systems. With new political anxieties and new scientific systems of understanding the body have come new digestive subjectivities.[34] Epigenetics, for instance, looks to notions of environmental and experiential exposure to understand digestion.[35] But we live today within a food system that presents food as an infinitely improvable, manipulable, economizable experiment, one that traces its origins to the time Darby first sliced his meat around 1870. Doing so created the conditions for a modern food system that routinely turns to technical fixes for complex economic and political problems, reduces food to its component parts, fetishizes consumer products, and prizes, above all, individual responsibility.

NOTES

1. Frederick William Pavy, *A Treatise on the Function of Digestion: Its Disorders, and Their Treatment* (London: Churchill, 1867), 274.

2. Chemical subspecialties underwent complex transformations in the course of the eighteenth and nineteenth centuries, dissolving traditional divisions such as animal and plant chemistry. Here, the designations of "chemist" and "physiologist" are used to indicate differences in methodology and to highlight the growing research enterprise of digestive ferments and artificial digestion, which explicitly positioned itself as a "physiological" alternative to "chemical" approaches to digestion and to food. They also reflect categories used by the historical actors examined in this chapter.

3. Sue Shephard, *Pickled, Potted, and Canned: How the Art and Science of Food Preserving Changed the World* (New York: Simon and Schuster, 2006), 34–40; Demet Güzey, *Food on Foot: A History of Eating on Trails and in the Wild* (London: Rowman & Littlefield, 2017), 135–152; Pierre Berton, *The Arctic Grail: The Quest for the North West Passage and the North Pole, 1818–1909* (Toronto: Anchor Canada, 2001).

4. Harriet Friedmann, "Feeding the Empire: The Pathologies of Globalized Agriculture," *Socialist Register* 41 (2005): 124–143.

5. The degree to which the Enclosure Acts themselves contributed to an increased agricultural output and represented a revolutionary rather than gradual shift in the relationship of small-scale agriculturalists to the land is a matter of debate. Readers can find the contours of that debate in such works as J. M. Neeson, *Commoners: Common Right, Enclosure and Social Change in England, 1700–1820* (Cambridge: Cambridge University Press, 1993); Robert C. Allen, "Tracking the Agricultural Revolution in England," *Economic History Review* 52, no. 2 (1999): 209–235; Robert C. Allen, *The British Industrial Revolution in Global Perspective* (Cambridge: Cambridge University Press, 2009), 57–79; Leigh Shaw-Taylor, "Parliamentary Enclosure and the Emergence of an English Agricultural Proletariat," *Journal of Economic History* 61, no. 3 (2001): 640–662.

6. Chris Cook, *The Routledge Companion to Britain in the Nineteenth Century, 1815–1914* (London: Routledge, 2005), 198; Richard Tames, *Economy and Society in 19th Century Britain* (London: Routledge, 2013), 21.

7. Boyd Hilton, *The Age of Atonement: The Influence of Evangelicalism on Social and Economic Thought, 1785–1865* (Oxford: Clarendon Press, 1988), 73–114.

8. Leslie A. Williams, *Daniel O'Connell, the British Press and the Irish Famine: Killing Remarks* (New York: Routledge, 2003).

9. John Bohstedt, *The Politics of Provisions: Food Riots, Moral Economy, and Market Transition in England, c. 1550–1850* (London: Ashgate, 2010).

10. Frederick Taylor, *In Memoriam Frederick William Pavy M.D., F.R.S., F.R.C.P* (London: Ash, 1913), 3; H. W. Bywaters, "Obituary Notices: Frederick William Pavy," *Biochemical Journal* 10, no. 1 (March 1, 1916): 1–4.

11. Pavy was appointed Medical Officer of Health of the district of St. Lukes and founded the Society of Medical Officers of Health. "Obituary: Frederick William Pavy," *Public Health* 25 (October 1, 1911): 72–73.

12. Pavy, *A Treatise on the Function of Digestion*; Frederick William Pavy, *A Treatise on Food and Dietetics* (Philadelphia: Lea, 1874).

13. Jeffrey Auerbach, *The Great Exhibition of 1851: A Nation on Display* (New Haven, CT: Yale University Press, 1999), 100.

14. I examine these processes in greater detail in my book manuscript "Wonder Foods: The Science and Commerce of Nutrition."

15. In other words, the imaginary imperial map of Britain's food supply included its formal and informal imperial territories. John Gallagher and Ronald Robinson, "The Imperialism of Free Trade," *Economic History Review* 6, no. 1 (1953): 1–15; P. J. Cain and A. G. Hopkins, *British Imperialism: 1688–2015*, 3rd ed. (London: Routledge, 2016).

16. Justus von Liebig, *Chemische Untersuchung über das Fleisch und seine Zubereitung zum Nahrungsmittel* (Heidelberg, Germany: C. F. Winter, 1847), 97.

17. Justus von Liebig, "Ueber die Bestandtheile der Flüssigkeiten des Fleisches," *Annalen der Chemie und Pharmacie* 62, no. 3 (1847): 257–368.

18. Justus von Liebig, *Familiar Letters on Chemistry: And Its Relation to Commerce, Physiology, and Agriculture*, trans. John Gardner (Philadelphia: Taylor and Walton, 1843), 27–30.

19. Emma Spary, *Feeding France: New Sciences of Food, 1760–1815* (Cambridge: Cambridge University Press, 2014), 90–91.

20. Ohad Parnes, "From Agents to Cells: Theodor Schwann's Research Notes of the Years 1835–1838," in *Reworking the Bench: Research Notebooks in the History of Science*, ed. F. L. Holmes, J. Renn, and Hans-Jörg Rheinberger (Dordrecht: Kluwer, 2003), 119–139.

21. Theodor Schwann, "Ueber das Wesen des Verdauungsprocesses," *Archiv für Anatomie, Physiologie und Wissenschaftliche Medicin* (1836): 90–138; Parnes, "From Agents to Cells"; Frederic Lawrence Holmes, *Claude Bernard and Animal Chemistry: The Emergence of a Scientist* (Cambridge, MA: Harvard University Press, 1974), 160–172; quotation from Schwann, "Ueber das Wesen des Verdauungsprocesses," 136.

22. I examine these developments in greater detail in "Wonder Foods."

23. Pavy, *A Treatise on the Function of Digestion*, 215.

24. Stephen Darby, *On Fluid Meat* (London: Churchill, 1870), 18–19.

25. William Roberts, "The Lumleian Lectures on the Digestive Ferments, and the Preparation and Use of Artificially Digested Food: Lecture III," *British Medical Journal* 1, nos. 1009–1010 (1880): 347.

26. Darby, *On Fluid Meat*, 16; Roberts, "The Lumleian Lectures on the Digestive Ferments," 347.

27. Darby, *On Fluid Meat*, title page.

28. Pavy, *A Treatise on the Function of Digestion*, 214–217.

29. Darby, *On Fluid Meat*, 19–20.

30. Pavy, *A Treatise on the Function of Digestion*, 574.

31. Edward Smith, *Dietaries for the Inmates of Workhouses: Report to the President of the Poor Law Board* (London: Eyre and Spottiswoode, 1866), 35.

32. Benjamin Ward Richardson, "The Skeleton in the National Cupboard," *The Asclepiad* 3 (1886): 218.

33. "Savory & Moore's Peptoniser," *The Medical Press and Circular*, October 3, 1883.

34. Hannah Landecker, "Being and Eating: Losing Grip on the Equation," *BioSocieties* 10, no. 2 (June 1, 2015): 253–258.

35. Hannah Landecker, "Food as Exposure: Nutritional Epigenetics and the New Metabolism," *Biosocieties* 6, no. 2 (June 2011): 167–194.

13

Modern Food as Substitute Food

ELLA EATON KELLOGG'S PROTOSE

FAKE MEAT AND THE GENDER POLITICS THAT MADE AMERICAN VEGETARIANISM MODERN

Adam Shprintzen

When you make a career out of studying the history of food, a trip to the grocery store can involve more than just checking items off a list. On occasion, a product, sign, or piece of marketing reminds me of something that I have researched, read, or wondered about. Sometimes grocery shopping elicits feelings of nostalgia for the moment I uncovered something in the archives—a photo, advertisement, or undiscovered letter. During a recent trip to the closest grocery store, amid dairy farms near my apartment in rural Northeast Pennsylvania, the history of vegetarian food—and of those who make and receive credit for it—was on my mind.

Modern food involved the creation of new products, often by women who never received credit for their work. Cruising through the frozen vegetarian food section, I was struck by the diversity of choice for such goods, especially for such a small store far from a major metropolitan center. Fake chicken nuggets, faux fish fillets, and phony barbecued pulled pork stared back at me from the freezer case, an abundance of vegetarian victuals. The brand names spoke to me. Gardenburger (will it sprout a burger if planted in soil?). The Impossible Burger (a name so confusing that it contradicts its very existence). And Beyond Meat (was meat a *limit?*). And wait . . . it "bleeds" beet juice? Something must be happening culturally and economically with these products because the same week that I made this grocery store visit, concerted efforts were being made by cattle ranchers to lobby state legislatures to ban the use of the word "meat" in marketing either lab-grown protein or plant-based meat substitutes.[1] Something must be happening, then, but it is nothing so new, at all. The history of modern vegetarianism is the history of substitute meat.

Uncovering the story of fake meat in America has occupied my thoughts and time for a dozen years. And it all started by stumbling upon a strange and unfamiliar word: Protose. I saw the name pop up repeatedly on online vegetarian message boards, but as a new vegetarian convert in the summer of 2008, I had no idea what it was. The name implied protein, but I had never seen such a product for sale.

I first tried to find the answer to what Protose was in recipes. The number of attempts to approximate Protose available online nearly matched the number of comments on vegetarian message boards decrying the product's demise. There were slight variations in each recipe, but the core ingredients were wheat gluten, ground nuts, and peanut butter. I decided to be bold and try to make my own version. But first some more research was necessary to find a more accurate recipe.

Scouring the historical record revealed that Protose's history stretched back to the late nineteenth century, a time when vegetarianism was growing significantly as a national movement in the United States. As the movement grew, so did its food choices, with the invention of meat substitutes like Protose. In late nineteenth-century America, all roads related to meat substitutes led to Battle Creek, Michigan and, in particular, to the experimental kitchen at the Battle Creek Sanitarium. By the mid-1880s, "the San," as it was called, had become the center of a growing phenomenon of health resorts in the United States, with the goal of curing any number of chronic illnesses. The San wasn't a hospital, though, but rather a vacation spot. A place to not only be treated for an illness, but also treated to time away from the busyness of everyday life.

An early photograph of the experimental kitchen neatly illustrates the how and why of the invention of Protose and other fake meats in Battle Creek. Kitchen workers stare with blank intensity at the camera. On the tables are a few dozen beakers; complex, interconnected Pasteur pipettes; and microscopes. Each member of the experimental kitchen staff donned a lab coat upon entering the room, rather than the white apron that was standard attire for the San's working kitchen that fed between 1,000 to

1,500 hungry guests and staff workers per day with recipes tested and created in the food laboratory.[2] The experimental kitchen could easily have been mistaken for a chemistry laboratory, and it served as both.

Ella Eaton Kellogg is central to this story. She was the wife of founder John Harvey Kellogg and took charge of this kitchen laboratory. She believed wholeheartedly in the importance of dietary reform and the benefits of scientific meatless dietetics.[3] The new reform cookery also needed to be delicious, moving beyond the stale bread and boiled vegetables of the past that often turned people away from a vegetarian diet. The modern, scientific, vegetarian kitchen needed to produce healthy *and* tasty dishes to keep food reformers happy and also bring new converts to the cause who would purchase the fake meats produced in Battle Creek.

For over forty years at the helm of the experimental kitchen at the Battle Creek Sanitarium, Ella Eaton Kellogg led the charge to reinvent and recreate vegetarianism as a culinary and social movement rooted in modern products. The vegetarianism that she helped create was driven by differing values from the movement's past, less concerned with the supposed dangers associated with the pleasures of food and more focused on food as a pathway toward personal, social, and economic reform. It's the origins of vegetarianism that most broadly interest me; it's a story that I have been telling in one form or another for more than a decade. And it's a story that keeps revealing new, previously hidden stories. To understand those origins means to understand meat substitutes. It's the buried story of Ella Eaton that interests me here because she is the forgotten character in the too-often masculine story of vegetarian history.

For such an important figure in the history of modern American food, remarkably few photographs of Ella Eaton Kellogg survive. Those that remain are from later in life and present a proud, somewhat frail-looking woman suffering under the dual forces of advanced age and chronic illness. Eaton Kellogg's drawn, fragile poses belie the spirit and drive that flowed through her work. Eaton Kellogg's friends, colleagues, and contemporaries noted this juxtaposition between image and personality in her final years, often speaking of the personal strength, wisdom, congenial personality, and scientific and medical acumen that defined her adult life.

Eaton Kellogg first became involved in health work when the newly professionalized medical field was first making space for women. Despite challenges from male practitioners, women began to receive formal training as nurses starting in 1873 with the establishment of nursing schools in New York, Boston, and New Haven, Connecticut. Women simultaneously entered the field as doctors thanks to the opening of the Women's Medical College of Pennsylvania in Philadelphia, which served as a training ground for women from around the world.

Eaton Kellogg did not follow a direct path toward a role in culinary innovation, though she displayed intellectual drive from an early age. She was born in Alfred Center, New York, in 1853, the daughter of Joseph Clarke Eaton and Hannah Sophia Eaton, and was educated into her teenage years at local schools. At the age of sixteen Eaton began attending Alfred University, where she was steeped in the teachings of the Seventh-Day Baptist movement that her family followed. Alfred was the second integrated college in the United States and was also remarkable for its coeducational studies.[4] At age nineteen Eaton became Alfred's youngest graduate.

In the summer of 1875, chance circumstances led to her role in shaping the trajectory of both the Battle Creek Sanitarium and American vegetarianism. Eaton and her sister traveled to Battle Creek to visit their aunt. Unfortunately, this trip coincided with a typhoid fever outbreak. Ella's sister became ill from the disease and, at the behest of her aunt, sought treatment at the growing Western Health Reform Institute—soon to be renamed the Battle Creek Sanitarium.[5]

The same year that Ella Eaton arrived in Battle Creek, a newly credentialed medical doctor named John Harvey Kellogg returned home to work as a staff doctor at the Western Health Reform Institute. They met at the bedside of a young patient who began to recover before relapsing into the grip of typhoid. The experience convinced Eaton to remain in Battle Creek and pursue a career in nursing.

By 1876 Eaton began working in the institute's newly created School of Hygiene while editing the institute's main publication, *Good Health*, a job that she would continue well into the twentieth century. Large changes were beginning at Battle Creek's premier health institute. In 1878 a new building was constructed on the grounds previously occupied by the Institute and was renamed the Battle Creek Sanitarium. Whereas the preceding management primarily emphasized water cure methods, the new organization delivered varied treatments. As these professional accomplishments grew, so did a fondness between Ella and John Harvey. The couple wed in February of 1879.[6] In the summer of 1883 the experimental kitchen at the Battle Creek Sanitarium opened its doors, with Ella Eaton Kellogg at the helm.

Eaton Kellogg's leadership role in the experimental kitchen acknowledged her sharp intellect. Still, this did not reflect a revolutionary challenge to gender roles or even a recognition of equality at the San, despite the responsibilities and seeming prominence. The experimental kitchen was established simultaneously with the rise of the domestic sciences as a movement and viable career option for middle- and upper-class women. The domestic sciences offered a career opportunity but also relied on an idealized notion of married women as moral guardians of the household, protecting both children and husbands. Among these important responsibilities were the ability and time to create healthy, flavorful, and satiating meals, a role that Eaton Kellogg filled at the San.

This ideology was fueled by a nativist desire to move middle-class households away from a reliance on immigrant domestics as home cooks. Eaton Kellogg exploited these fears of recent immigrants, explaining once that the new scientific cookery was necessary to encourage homemakers because in many middle-class homes, "cooking is trusted to the ignorant class of persons having no knowledge whatever of the scientific principles involved in this most important and practical of arts."[7]

The experimental kitchen appeared at a moment of transition for the vegetarian movement in the United States. For the first half of the nineteenth century, vegetarians connected with a variety of radical social reform movements. The group viewed their diet as the center of a total reform lifestyle, with the power to transform American society by supporting such movements as women's rights, pacifism, and abolitionism. A focus on abolitionism led movement vegetarianism to dissipate during the late 1850s. As a result, the postbellum years allowed vegetarianism to reinvent itself through a new movement. But what would be at the center of this redefinition?

Eaton Kellogg viewed the experimental kitchen as a key component in the process of vegetarianism as a new scientific cookery movement. The scientific kitchen focused on the development of cereals. Granola's eventual popularity seemed to prove the effectiveness of scientific experimentation to improve food quality and convenience. Granula was originally invented at the Our Home resort in Dansville, New York, as a harsh mixture of Graham flour, water, and grains rolled into a dough, baked in a brick oven, and crumbled into barely digestible pieces. The cereal's dryness made it difficult to chew and inconvenient to prepare: in order to be soft enough to eat, it needed to soak in milk overnight in an icebox.

Granula was reinvented in the experimental kitchen under Ella Eaton Kellogg's watchful eye. Workers put together differing combinations of grains and cereals, attempting to craft a product that was healthy, flavorful, and convenient. The cereal was renamed granola to avoid copyright infringement. Rather than dense Graham flour, the final recipe for the San's granola included ground zwieback—a sweet, crusty rusk of Central European origin served and sold by the San—mixed with grain. The new granola did not need to soak in order to be edible and was more palatable due to its sweetness.[8] As Michael S. Kideckel discusses in chapter 7, many of the grain products developed in the experimental kitchen were later among those marketed as breakfast cereal, a key food in a processed food supply branded as ethical.

Rethinking already existing foods helped ensure that consumers at the San and at home did not view scientific foods as medicinal. Eaton Kellogg followed a precise method when developing and marketing vegetarian foods. The first step was to identify and then replace what she deemed to be bad foods, meats and dairy in particular. Once she categorized these foods, she worked with experimental kitchen staff to invent replacements through experimentation. Eaton Kellogg took particular interest in figuring out the "blood making" capabilities of different foods, a nod to the protein in nut-based products. Eaton Kellogg then took that information to the laboratory, where she cooked new dishes based on her own research alongside advice from experimental kitchen workers. Her cooking process mirrored that of scientific experimentation, with a basic hypothesis of what would be flavorful and healthful, and subsequent adjustments based on the results. Many of these dishes ended up being popularized in cookbooks authored by Eaton Kellogg.

By the late 1880s Eaton Kellogg began turning her attention to the use of nuts, particularly as a source of protein for both the San's residents and vegetarian consumers around the country. Advertisements and articles from Battle Creek emphasized that nuts had protein qualities similar to meat without ethical or health concerns. The accuracy of these claims was less important than the ways that the experimental kitchen created a sense of scientific authority and dietary expertise to promote a new type of vegetarianism.

This brings us to Protose, the fake meat that dedicated vegetarians would mourn years later. Early in 1896 an official from the US Department of Agriculture (USDA) wrote to the San, interested in supporting the development of what he described as "a scientifically prepared plant product affording all the essential nutrient qualities of beef or mutton . . . a food product which might be safely employed as an alternative for meats," given the growing price of meat and his fear of a possible meat shortage.[9] Eaton Kellogg deftly engineered a solution that many facing pandemic-revealed limitations in the meat supply chain would envy today and created the first wave of American fake meats.

Within a year she oversaw the creation of Granose, a wheat-based biscuit that could be utilized as a faux fillet of beef. Granose was the first of nine meat substitutes created by the experimental kitchen, expanding vegetarian food choice in both America and around the world. Meat substitutes developed in the experimental kitchen not only changed culinary choices for vegetarians but also shifted the goals of the movement. Protose: its taste, its texture, and its popularity, accelerated this change.

The story of Protose explains how Ella Eaton Kellogg changed the vegetarian movement's ideals through food production, consumption, and marketing. Protose was crafted through rigorous, repeated testing in the experimental kitchen, with Battle Creek food scientists eventually settling on a combination of wheat gluten, cereals, and peanut butter. Protose had the qualities of canned, chopped "vegetable meat," a culinary canvas for vegetarian epicures to shape, cook, and consume in as many ways as the imagination could concoct.[10] The vegetable meat looked, felt, and smelled like meat, and it had the added benefit of being cheaper. More than a century later, plant-based meats are being marketed based on the same supposed qualities.

A one-pound can of Protose sold for thirty cents in 1912, whereas a one-pound can of canned meat sold for thirty-five cents in the same year.[11]

The federal government was already debating how to classify fake meats as early as 1910. Should they be classified as canned meats or canned vegetables? The modern ranchers who are threatened by the existence of fake meat products like Beyond Meat may be torn by the historical precedent. The government did not classify fake meat as meat, but it also allowed for the use of the word "meat" for vegetable-based products. In the process a new category regulating the shipping of "fake meats" was created.

Despite the attention given to Protose, Ella Eaton Kellogg did not receive direct accolades at the time of the product's spread around the country. Although the USDA celebrated the vegetable meat as "widely known" and "manufactured and used in the leading civilized countries in the world," the assistant secretary writing the report, Charles Dabney, ignored Eaton Kellogg's central role in his telling of the development of fake meats.[12] Instead, he explained, "*Dr.* Kellogg (J. H.) undertook a research which has resulted in the development of a number of meat-like products."[13] In his diaries Dabney credited John Harvey Kellogg with this innovation. Ella Eaton Kellogg was unceremoniously cut out of the account.

And yet, Eaton Kellogg's role in spreading the gospel of vegetarian eating did not end at the experimental kitchen. She stood at the center of efforts to market Protose and other vegetable meats as well as the ways to prepare these products. Eaton Kellogg authored cookbooks that spread awareness of the products and their preparation while emphasizing the values of the new vegetarianism: social success, economic advancement, and reform of the self. Early descriptions of Protose that she crafted explained that it was "one of the latest and greatest triumphs of modern discovery . . . so closely resembling meat in appearance, flavor, and texture as almost to deceive an epicure."[14]

The scientific food reform movement that Eaton Kellogg helped grow utilized the printed word to spread its message and methods through domestic advice in magazines, journals, and cookbooks. In one cookbook Eaton Kellogg explained that the scientific kitchen brought "order from out the confusion of mixtures and messes . . . with the same certainty with which the law of gravity rules the planets."[15] Because scientific experimentation served as the backbone of the new cookery, its results could not be denied or disputed. In essence, dietary reformers sought to ward off doubters by claiming measurable expertise, a particularly important strategy for Eaton Kellogg in arguing in favor of vegetarianism to the meat-consuming masses.

Science was necessary in all kitchens, not just at Battle Creek, Eaton Kellogg explained, because of its fundamental importance to human health, industriousness, and morality. Therefore, food preparation was "deserving of the most careful consideration. It should be studied as a science, to enable us to choose such materials as are best adapted to our needs."[16] Of course, there was a built-in, significant expense with this new vegetarian cooking. A home scientific kitchen necessitated far more cooking equipment than the working classes could ever afford. She promoted the idea that a home kitchen should be a "workshop" that needed to include such kitchen conveniences as a vegetable press, a lemon drill, a handy waiter (a table on caster wheels used to bring food to the table or to help clear dishes), a separate kneading table to make fresh bread, and a wall cabinet filled with fifteen different cooking utensils and jars. Eaton Kellogg's intent was clear: it was necessary to target vegetarianism toward the middle and upper classes, those who could afford an extensive home kitchen, because they would be the ones who would buy Battle Creek products through mail order.

Unsurprisingly, it was foods often found on the Battle Creek menu—nuts and legumes in particular—that Eaton Kellogg deemed as most advantageous to home cooks in her writings. Eaton Kellogg's first foray into directly marketing meat substitutes in a cookbook occurred soon after the development of products such as Nuttose and Protose. In 1897's *Every-Day Dishes and Every-Day Work*, she explained that the recipes were the result of "work carried on in the Experimental Kitchen of the Battle Creek Sanitarium."[17] The cookbook included an entire chapter entitled "The Battle Creek Sanitarium Health Food Company's Products." Granola was highlighted as versatile—for use as a breakfast cereal or breadcrumbs for savory vegetable dishes. A vegetable roast could be created with rehydrated lentils, nut butter, and crystal wheat—a concentrated wheat grain. Nuttose was the first meat substitute explicitly marketed as such from the experimental kitchen. It could be used similarly to "the various forms of flesh food." Eaton Kellogg reported that Nuttose "so perfectly resembles meat in appearance and flavor . . . that many persons find it difficult to distinguish the difference." Recipes included a stewed Nuttose with tomatoes, Nuttose with green vegetables, and a Nuttose hash.[18] The goal of these dishes, she explained, was to create healthy, successful bodies prepared for social success, or, as Eaton Kellogg explained in a nod to the growing eugenics movement and its racist hierarchies, the "strongest, and most enduring races."[19]

The unwriting of Eaton Kellogg's lead role early on is surprising given that in 1904, just a few years later, she placed meat substitutes at the forefront of her book *Healthful Cookery*. The early pages of the cookbook represented a new approach toward meat by vegetarians. Eaton Kellogg explained that meat had beneficial properties in terms of both protein and fat content. Instead of attacking all properties of meat, she explained that fears of tainted meat ensured that the "important food elements" that meat provided should be replaced with meat substitutes. The new marketing ploy was a clear attempt at selling the San's products not just to dedicated vegetarians but also the carnivorously inclined to ensure commercial success. I see a continuity here with the marketing of modern meat substitutes, targeting not just the vegetarian and vegan faithful but also so-called flexitarians and individuals looking to decrease their meat consumption for economic, health, or environmental reasons.[20]

Eaton Kellogg described Protose as "the perfect substitute for flesh food," and a section of the cookbook referred to the San's products as "Flesh-Food Substitutes."[21] The cookbook included over 100 different preparations of Protose, ranging from a Protose loaf with the addition of bread crumbs, salt, and sage, a Protose fricassee stewed with tomatoes and peanut butter, and a mock hamburger steak including Protose, eggs, and Granose to bind everything together.[22] These recipes were all designed and tested in the experimental kitchen. The cookbook positioned meat substitutes as flavorful, versatile, and maximizing the nutritional values associated with meat. Protose could be morphed into any shape, baked, broiled, breaded—just like the mass-marketed meat substitutes of today.

The method of marketing Protose and other meat substitutes surely expanded the potential market for the San's products. The number of Americans eating vegetarian foods increased, and many for the first time came to see vegetarianism as a viable choice. There was also a financial payoff for Battle Creek. Protose went nationwide by the turn of the twentieth century and worldwide by the end of its first decade, found as far

away from Battle Creek as health food stores in Australia. By 1914, shipments of more than 144,000 pounds made their way across the United
States.[23] During that same year over 33,000 pounds of Nuttolene (a vegetarian loaf made primarily of peanuts) were sold across the country, along
with another 6,000 pounds of Nuttose.[24] The amount of meat substitutes
consumed in the United States was tiny compared to overall meat consumption. But by 1914, meat substitutes created in Ella Eaton Kellogg's
experimental kitchen at the San were well known, widely consumed, and
profitable.[25] For the first time, vegetarianism in America was a commercial success, driven primarily by profit rather than ethics. Vegetarianism
was reinvented with new products, and Ella Eaton Kellogg stood at the
center of this significant change.

As a food product, Protose was a revolutionary shift away from the basic
preparations of vegetables that defined vegetarian diets in America through
most of the nineteenth century while also ushering in a new vegetarian
movement. Tasting Protose for the first time must have been quite the
sensory experience for a long-term vegetarian in late nineteenth-century
America. The product was the culmination of more than a decade's work
in the Battle Creek Sanitarium's experimental kitchen and was a significant
change in the nature of vegetarian foods available. The plain, basic fare
that fueled the social reformist crusades of radical vegetarians of the first
half of the nineteenth century gave way to new, so-called meat substitutes
that provided sustenance for individuals competing in a highly competitive social and economic world. For new converts to the cause, drawn by
promises of economic and social promotion through the consumption of
fake meats, Protose was meaty enough to make a change in diet seem less
dramatic. Nuts and cereals bridged the gap between the boiled vegetables
and bran bread of the 1850s to Protose of the 1890s, which was marketed
as looking, tasting, smelling, and feeling like meat.

The fuller story of Protose's history recently led me to try a bout of historical recreation. I decided to make Ella Eaton Kellogg's recipe for a Protose
loaf that was introduced in the 1890s. As a raw mixture, the Protose visu-
ally resembled the consistency of the sort of canned meats that Anna Zeide
describes in chapter 10. To the touch, the "raw" Protose also felt like canned
meat. The fake meat's texture was creamy thanks to the introduction of

peanut butter, but also had a canned meat–like thickness because of the amount of ground nuts, grains, and cereals. The "raw" product was salt-forward given a lack of spices and seasonings, other than a little bit of sage. Once I baked the Protose in a casserole dish, the loaf's taste and consistency changed dramatically. I was struck by the product's crust, which resembled a piece of broiled meat, a common preparation of Protose that was promoted in Battle Creek cookbooks and frequently prepared at the San for guests. When baked, the Protose loaf—in the tritest food cliché possible—tasted just like chicken, a quality emphasized by the earliest advertisements promoted by the San's mail order company. To my modern palate, with decades of experience eating heavily spiced fake meats, some made of nuts, others from soy and other legumes, Protose tasted bland. But to Ella Eaton Kellogg, I imagine the first taste of Protose was a bit of a revelation in experiencing the possibilities of how meat could be mimicked. Consumers apparently agreed. But Eaton Kellogg's and Protose's legacy survived only for a limited time in the public understanding.

In Ella Eaton Kellogg's later years, health issues connected to a childhood bout with scarlet fever began to affect her hearing. She learned to become adept at lip reading and continued her work editing *Good Health* while leading efforts in the experimental kitchen. Poor health, however, continued to plague her in the late 1910s as she became stricken with cancer. Ella Eaton Kellogg passed away in June of 1920 at the age of 67. Newspaper accounts of her death perpetuated the gendered masking of her pioneering role in the history of domestic sciences and scientific cookery, giving equal play to her role as a cookbook author with her status as spouse to John Harvey Kellogg. I have yet to find a single obituary that mentioned her work in the experimental kitchen.[26] Digging deeper into this story has also forced me to come to terms with my own role in marginalizing Eaton Kellogg's role at Battle Creek, something I overlooked and underplayed in both my book on vegetarianism history and a journal article on the subject of the San. Historians interpret the past. Unfortunately, without a concerted effort we tend to interpret through our own gendered lens. Commercialized vegetarianism had room for Ella Eaton Kellogg, but only as a domestic advisor. Likewise, our collective historical memory of Ella Eaton Kellogg has not previously allowed room for her transformative role as an inventor, innovator, and food scientist.

Protose, despite its long-term popularity among vegetarians, had an expiration date as well. Even though it lasted much longer than its inventor, it also could not escape the implications of corporatized food. Protose remained popular throughout the twentieth century after Eaton Kellogg's death, marketed as a meat substitute for vegetarians by Worthington Foods, a Seventh Day Adventist company that was the first to market frozen vegetarian meals in supermarkets in the United States. However, in 1999 a large international food-producing conglomerate purchased Worthington Foods. The Kellogg corporate behemoth—founded by Ella Eaton Kellogg's brother-in-law, Will Keith, in 1906—bought Worthington Foods in 1999 and discontinued Protose a year later.

The brand was bought and squashed in order to knock out competition with Kellogg's growing Morningstar Farms brand of fake meats, which included fake chicken cutlets, veggie burgers, and faux sausages. But the story continues to evolve for fake meat and the Kellogg name. In 2019, Kellogg's announced the future release of its Incogmeato brand, to be available on the shelves of large grocery store chains such as Kroger, Safeway, Albertsons, and Weis. Incogmeato is promised to "Make Meat Jealous" as the product can "Trick your taste buds with 100% plant-based protein that looks, cooks & tastes like meat." More than a hundred years later Kellogg's is unwittingly looking to the past for inspiration, echoing the marketing of Protose, which was purported to look, taste, and smell like meat. The question today is: Will vegetarians and other consumers of fake meat choose to utilize the diet to help create a better world, or only to fashion a "better" self? Almost a century after Protose's invention by a forgotten food innovator, commercialized vegetarianism continues to foster a community of likeminded eaters consuming meatless cutlets, loafs, and steaks, fueled by the promise of vegetarian-fed, socially successful, and economically advancing bodies.

NOTES

1. Nathaniel Popper, "You Call That Meat? Not So Fast, Cattle Ranchers Say," *New York Times*, February 9, 2019, https://www.nytimes.com/2019/02/09/technology/meat-veggie-burgers-lab-produced.html.

2. Descriptions of work uniforms come from photographs of each of these rooms at the San. See John Harvey Kellogg Papers (hereafter JHKP), Bentley Historical Library,

University of Michigan, Photographs, Battle Creek Sanitarium, Various Sanitarium Events and Activities (ca. 1910–1943), 16, 22. The estimate of the number of diners at the San comes directly from John Harvey Kellogg and seems to include all meals prepared in a given day for both staff and residents. See John Harvey Kellogg, *In Memoriam: Ella Eaton Kellogg* (Battle Creek, MI: Battle Creek Publishing Company, 1920), 20. The number seems accurate. For example, in 1911—the closest annual report to Ella Kellogg's year of death—the San welcomed 5,035 patients through the year. See *Annual Report of the Battle Creek Sanitarium and Hospital* (Battle Creek, MI: The Sanitarium, 1912), 31.

3. Ella Eaton Kellogg, *Science in the Kitchen* (Chicago: Modern Medicine, 1893), 12.

4. For more on Alfred University's nineteenth-century history, see Susan Rumsey Strong, *Thought Knows No Sex: Women's Rights at Alfred University* (Albany: State University of New York Press, 2008).

5. Kellogg, *In Memoriam*, 7.

6. Kellogg, *In Memoriam*, 8.

7. Kellogg, *Science in the Kitchen*, 46.

8. Andrew F. Smith, *Eating History: Thirty Turning Points in the Making of American Cuisine* (New York: Columbia University Press, 2009), 142; John Harvey Kellogg, "The Natural Diet of Man," JHKP, Box 3, Folder 23 (April 16, 1900), 1–2.

9. "Notes and Memoranda, Diet," JHKP, Box 8, Folder 5.

10. Lenna Frances Cooper, *The New Cookery: A Book of Recipes, Most of Which Are in Use at the Battle Creek Sanitarium* (Battle Creek, MI: The Good Health Publishing Co., 1913), 307.

11. Cooper, *The New Cookery*.

12. Cooper, *The New Cookery*.

13. "Notes and Memoranda, Diet."

14. Sanitas Nut Company, *The Nut Cracker* 1, no. 1 (July 1900): 11, 18.

15. Kellogg, *Science in the Kitchen*, 4.

16. Kellogg, *Science in the Kitchen*, 1.

17. Ella Eaton Kellogg, *Every-Day Dishes and Every-Day Work: A Collection of Choice Recipes for Preparing Foods, with Special Reference to Health* (Battle Creek, MI: Modern Medicine Publishing Company, 1897), 3.

18. Kellogg, *Every-Day Dishes and Every-Day Work*, 145–148.

19. Kellogg, *Every-Day Dishes and Every-Day Work*, 18.

20. Ella Eaton Kellogg, *Healthful Cookery: A Collection of Choice Recipes for Preparing Foods* (Battle Creek, MI: Modern Medicine Publishing Company, 1904), 72–75.

21. Kellogg, *Healthful Cookery*, 91.

22. Kellogg, *Healthful Cookery*, 75, 78, 92.

23. Herbert Confield Lust, *United States Interstate Commerce Commission, Supplemental Digest of Decisions under the Interstate Commerce Act* (Chicago: Traffic Law Book

Company, 1914), 154; Economic Research Service (ERS), U.S. Department of Agriculture (USDA), *Food Availability (Per Capita) Data System,* http://www.ers.usda.gov /Data/FoodConsumption.

24. United States Interstate Commerce Commission, *Supplemental Digest of Decisions under the Interstate Commerce Act* (Chicago: Traffic Law Book Company, 1914), 154.

25. As a comparison, the average annual consumption of meat per capita in the United States in 1912 equaled 145 pounds. See Roger Horowitz, *Putting Meat on the American Table* (Baltimore: Johns Hopkins University Press, 2006), 12.

26. See, for example, "Ella Eaton Kellogg," *Owensboro Messenger,* June 16, 1920, 1; "Deaths in Michigan," *Times Herald* (Port Huron, MI), June 15, 1920, 14; and "Mrs. Kellogg, Who Aided Many Children, Is Dead," *Chicago Tribune,* June 15, 1920, 19.

14

Modern Food Is Not Your Pet

MARIETTA'S LAMB

Amrys O. Williams

Fourteen-year-old Marietta Hamilton knelt with her lamb in the snow of a Great Falls, Montana, January. She was learning to become a modern eater, someone who understood food through systematic and rational metrics. I came across this picture of her (figure 14.1) as part of my research for a larger project on the ways agricultural education, and 4-H in particular, helped to shape that "modern" eater. You can learn a lot from a photo of a girl and her lamb.

Marietta's straight brown hair was cut short in the 1930s style: parted on the side, it framed her face underneath a pale, jaunty hat. Her fur-trimmed, belted wool coat protected her from the wind whipping through the streets as she faced the photographer. Beside her, warm in his own woolly coat, was the lamb she had spent the past year raising. Her right hand gently directed his gaze toward the camera. The snow creaked under her leather boots until both she and her lamb were still. The shutter clicked.[1]

Around her swirled the activity of the First Annual Montana 4-H Fat Lamb Show. Kids from across the state chattered as they lined up with their prizewinning lambs to have their pictures taken by the local portrait studio. The scent of tobacco mixed with the smells of hay and animals on the winter air. Adults in suits and overcoats milled about the edges, smoking and talking about sheep breeding and the price of lamb. Beyond them stretched the stock pens filled with a chorus of bleating lambs raised by

14.1 Marietta Hamilton poses with her prizewinning lamb at the Montana Wool Growers' Association Fat Lamb Show, Great Falls, January 1933. (D. P. Thurber, "Annual Report of the Extension Agent for the Sun River Irrigation Project" [unpublished report, 1933], 39.)

4-H club boys and girls throughout Montana. Marietta had brought her best lambs here to compete, with the hope that they would fetch both a prize and a good price. Her efforts weren't just fun and games. It was the start of 1933, and the Great Depression was deepening. It spelled hard times for farm families like hers. The food they raised was important, for themselves and the country as a whole—and, though it didn't look like it just yet, Marietta's lamb was first and foremost food.[2]

I connected with Marietta immediately. When I was fourteen, some sixty years later and thousands of miles away, I wanted to be her. Just like Marietta, I had joined a 4-H sheep club, full of dreams of being a farmer. I wanted to be one of those kids who went to the fair to show their stock, who knew how to handle animals with ease. But I lived in the city, and raising a sheep in our small backyard would have been impossible. So I made pies and picked strawberries and learned about ovine anatomy and nutrition at 4-H club meetings, but I never showed a sheep at the fair.

Marietta's story was thus both familiar to me and new, an encapsulation of all my youthful dreams of rural life and an indication of how that life was changing in the early twentieth century. The photographer had captured Marietta and her lamb in a pose I had seen countless times: a 4-H member with her project, showing how club work was making American agriculture and its practitioners better and more productive. But the next photo I stumbled across was new to me, and it made me stop and think. It showed the front window of the S&B Grocery in Great Falls in the days after the Fat Lamb Show, decked out with an attractive display of eggs, tomatoes, celery, and coffee cans stacked up on the sides (figure 14.2). Right in the middle was a fresh lamb carcass, chest split open, garnished with parsley, a few root vegetables spilling abundantly from the cavity like a cornucopia. A hand-lettered sign advertised "4-H Club Grand Champion Lamb," and the name of the boy who had raised it—Marietta's neighbor, Bernard Hatcher. To 4-H club members, this tableau was a sign of a profitable sale and a job well done. But to me, and to the passersby of Great Falls, it was emphatically a picture of food. I could envision the next scene: a Great Falls family, seated round the dinner table, tucking in to a hearty meal of roast Montana lamb.[3]

Before these lambs became a meal, they were a yearlong lesson in eating: for the lambs, for Marietta and Bernard, and for the nearly 1 million other 4-H club members who were raising poultry and dairy calves, cultivating acres of corn, and planting vegetable gardens across the United States in the early 1930s. At a time of enormous change in American agriculture, these young people were participating in the government's attempt to harmonize industrial modernity and rural life. By using science to make that life more rewarding, the agrarian foundation of the nation could endure. This effort reached into barns and farmhouses alike, grounding research findings in the daily activities of farm families: feeding livestock, raising crops, growing gardens, cooking meals. A chain of eating lay beneath it all.[4]

In our current age of distant factory farms, long-haul trucks, and supermarkets, urban consumers try to learn more about where their food comes from. But Marietta and her community were intimately familiar with the farm, their food's source. That means when 4-H—and the US Department of Agriculture (USDA), the federal agency behind the clubs—sought to

14.2 A 4-H lamb carcass formed the centerpiece of the S&B Grocery's window display just after the Fat Lamb Show in January 1933. (D. P. Thurber, "Annual Report of the Extension Agent for the Sun River Irrigation Project" [unpublished report, 1933], 43.)

convey food and nutrition education, it wasn't because their rural targets didn't understand food's origins. Instead, they were trying to link nutrition on the farm with the nutrition of the farm family. They wanted to transfer the knowledge of how to scientifically feed animals to the scientific, modern feeding of humans. Marietta came to feed her sheep in a modern way, and in so doing she herself became a modern eater.

Home for Marietta was a farm on Montana's Fairfield Bench, a wide plateau above the valley of the Sun River that flowed into the Missouri at Great Falls to the southeast. Marietta and her parents were among the new families to arrive there in 1931 as part of the Sun River Irrigation Project. This massive federal reclamation effort brought water to 91,000 acres of dry land to be farmed according to modern methods by folks like the Hamiltons. Late in the summer of 1932, Marietta spotted a car kicking up dust along the road to her family's place. She recognized the man in a suit who stepped out as Mr. Thurber, Sun River's extension agent. He had offered her family advice on which crops did best on the newly watered lands, helped them purchase good seed, and encouraged them to start raising sheep.

Thurber had a proposal. He wanted to see if the Hamiltons would be interested in fattening some of their lambs to market this winter—and if Marietta would agree to take charge of it and show twenty-five lambs at the Wool Growers' convention in January. She'd take home the money she made on prizes and at the auction. Marietta must have been interested and her parents amenable because she was one of three kids—and the only girl—to join the first 4-H sheep club on the Sun River Project.[5]

For Thurber, Marietta's participation in the lamb project was one piece of a larger puzzle of agricultural planning and development in Montana and across the nation, one focused on maintaining an agrarian foothold for farm families amid the urbanization of the machine age. Concerned about agrarian decline, the USDA and the state extension services tried to help farmers weather the economic turmoil of the 1930s through advice and assistance, and they saw modern irrigated homesteads like the Hamiltons' as models for the rest of rural America. Sun River farmers produced abundant forage and feed crops on their irrigated lands, including alfalfa, wheat, oats, and barley. But the planners were also concerned about what would happen if they produced too much—more than they could profitably sell in an uncertain market. One solution—as in chapter 1 of this collection, for example—was to develop global markets for the surplus and cultivate new consumers. In 1930s Montana, planners took a more regional approach. By fattening livestock on the excess feed, farm families like the Hamiltons could sell profitable meat as well as grain and hay. The 4-H lamb feeding project was about turning a potentially price-depressing surplus into a marketable, high-value farm product. It was part of a diversified program for Montana agriculture that would benefit farmers, farm organizations, and the state. For Marietta to learn to be a modern eater, she would have to navigate the terms of Thurber's world, where measurement, quantification, and scientific record keeping defined farming.[6]

We often talk today about the relationships between local, individual action and global, collective movements. This local-global tension isn't new, though the forms of the relationships depend on the era. Several others in this anthology speak to macro-level versions of it, as in Lisa Haushofer's discussion of the imperial tenets of British digestion (chapter 12) or Jeffrey Pilcher's tale of local Czech pilsner as the first global beer (chapter 4). Marietta's lamb gives us a more microcosmic view and shows us how these tensions played out in the daily lives of the people who produced food. Thurber was an agent of the federal bureaucracy; Marietta was a teenager finding her bearings on a wide plateau in Big Sky country.

The local and the global came together in her sheep. A new breed developed by the USDA for the mountain West, the Hamiltons' flock symbolized the alliance between agricultural research and rural development. By joining the fat lamb project, Marietta would carry that alliance forward, as would hundreds of thousands of others like her. Imagine all those Mariettas, lined up with their lambs and calves and hogs, mugging for the camera at every fair and livestock show from Maine to California. Take all those pictures and put them in a flipbook, and you can start to see the filmstrip of the transformation of American agriculture, the creation of modern food.[7]

The image of 4-H clubs today, if people know them, remains one of rural livelihood—which was one reason I was attracted to them as a youngster, even though it was hard for me to have a sheep during my urban childhood. In Marietta's time, 4-H was establishing itself as a training program for all aspects of rural life, from the field to the farmhouse. Food became particularly important during the interwar years as doctors, nurses, and a growing cadre of nutrition specialists grew alarmed at an emerging crisis: the poor health of the very people who produced the nation's food. Servicemen enlisting in World War I had all received physical examinations, giving the federal government an unprecedented survey of the general health of the American population. They found widespread nutritional deficiencies and other problems resulting from poor or inadequate diets. The situation was particularly dire in rural districts—a fact which came as a shock to many Americans, who assumed that the countryside was a more healthful place than the crowded industrial city. Home econo-mists and nutrition specialists began a crusade to restore and improve the health of the farm families who fed and clothed the nation.

To wage their battle, they looked to what the Extension Service had done to improve crop production and animal husbandry on the farm through the work of agents like Thurber. Since 1914, agricultural spe-cialists had used 4-H clubs to teach methods of farming and livestock raising based on the latest scientific research. This widespread effort to grow better plants, animals, and people was a food reform movement that transcended species, involving chemists, agricultural scientists, and home economists. And the Extension Service translated these findings into guidelines for farm families to follow.[8]

Farm women and girls like Marietta were key players in this effort. They raised and fed babies and children, selected and purchased food, prepared the family's meals, and were often in charge of growing what food the farm produced for home use. To a great extent they determined how well the family ate. If they grew fresh vegetables, canned their produce for winter use, and raised enough chickens and cows to provide the family's eggs and milk—and knew how to prepare balanced, tasty meals from these offerings—they could ensure that their children grew up healthy and strong.

On Sun River, many girls joined the 4-H sewing and cooking clubs that focused on these kinds of home-economic skills. But Marietta apparently saw herself as more than a farmer's wife in training. It told in her every-day outfits. Rather than the cotton dresses common to 4-H clothing club members, she wore wide dungarees rolled at the ankles to fit her young frame and simple blouses to keep her cool under the hot sun of a Mon-tana summer as she worked in the pastures and the sheep pen. And while she didn't enroll in a cooking club project, she learned a great deal about food from her lambs.[9]

This was because nutrition reformers took a two-pronged approach to improving rural diets. They focused on the ingredients side by encourag-ing women and girls to keep gardens and livestock and teaching them to preserve food for year-round consumption. They also grappled with the meal side by teaching them the basics of nutrition, providing menus and recipes for healthy and tasty dishes, and suggesting how to make meals attractive and pleasant as well as nourishing. In extension work in agriculture and home economics, a farm family's food was the crucial link between agricultural and social reform.[10]

By the early decades of the twentieth century, chemists and home economists had brought terms like carbohydrates, fat, protein, and vitamins and minerals into the kitchen and taught Americans that food was a means of delivering energy to the body in calories, endowing it with the building blocks for muscle, bone, and other tissues. By thinking of food *as* nutrients, people could equate vastly different foodstuffs that had similar nutritional or energy profiles.[11]

Meanwhile, rural people were taking up nutritional science at least as systematically as their urban counterparts—only not, at first, for themselves. Rather, farmers were enthusiastically applying nutritional principles to their crops and livestock, whose "diets" were much simpler to control and rationalize. The sheep came first; the humans followed. 4-H projects like Marietta's were one of the more crucial yet underappreciated routes that nutritional science took into the homes, kitchens, and daily lives of rural Americans.[12]

If Marietta's experiences give us a snapshot of the making of a modern, rational eater, then the details of her 4-H education are illuminating. The USDA can have big ideas, but it's the mundane labor of daily tasks like Marietta's 4-H fat lamb regimen that carries those plans out. How do you spread new food habits, for instance, whether for lambs or girls? If the modern eater is one of facts, figures, and systematic diets, how do you develop that on the farm? For 4-H, the answer was the record book.

When I was initially researching Depression-era 4-H, I focused on county agent reports like Thurber's. But project bulletins and record books were what kids like Marietta used to carry out their projects, and these unlock one of the more important parts of club activity: record-keeping. Record keeping was scientific and economic. It focused on measurement and quantification, in terms of time, supplies, and money. It encouraged club members to get the most out of what they put in, and it encouraged them to reflect on what they had done and make improvements as they went. The payout for good records was a good showing at the state fair— and that meant both accolades and cash in hand.[13]

In Marietta's case, the "Record Book for 4-H Boys' and Girls' Livestock Clubs" encouraged her to think about food in truly rational terms. "Ration" and "rational" may conjure up different images—scarce supplies and war-time shortages on the one hand, passionless calculating on the other—but their root is the same: a reckoning. Keeping records on inputs and outputs helped transform farming and eating. It turned feed into a ration and feed-ing into a rational act—first for the sheep, then for the people.

Marietta would have begun filling out her record book on the day in late September or early October of 1932 when she selected her twenty-five lambs from the family flock. This was an important enough event that county agent Thurber and a man from the agricultural college drove all the way out to the farm to observe her work. I could see Marietta, leaning on the fence, thumbing through her 4-H bulletins, checking off the lambs' qualities in her head and deciding which ones would be good "feeders" and gain weight fast according to what she had read. Removing her record book from the hip pocket of her dungarees, she wrote down the lambs' names and dates of birth before bringing each one to the scale for weighing. The two men signed her record book, certifying that the weights were accurate. They then looked on as Marietta added her signa-ture to the document. Her fat lamb project had officially begun.[14]

From here, Marietta would complete her record book. Picture her filling in the blanks with a stubby, knife-sharpened pencil or a school fountain pen. On the first page she wrote her name, age, county, and community. *Marietta Hamilton, 14, Teton County, Fairfield.* She noted the type of club and the year, the club's name and its leader. *Fat lamb, 1932, Fairfield County-Wide Club, D. P. Thurber.* She then read over the instructions and pointers, which provided tips on better feeding under the watchwords "Make every pound of feed yield a profit." She copied the names of the lambs and their initial weights onto the weight gain table and turned the page.

Here was the important part: her feeding record (figure 14.3 shows a blank copy of the page). She kept careful track of what she fed her lambs, when she fed them, and how much it cost. Her goal was to get each of her lambs from the sixty-odd pounds they weighed in October to around ninety pounds by the time of the January lamb show, ninety days away. This meant gaining a third of a pound each day. Marietta needed to provide the food to make that happen, and to do so she would hew

Kind of Club..No. of Demonstration............................

Breed ....

Purebred, grade or crossbred..

Date record started...Date record ended...

Table I—Animals Used

Record initial and final weights and values of animals here. Weights of breeding animals as well as those in feeding demonstrations should be kept in order to have a record of their growth.

Name and Registration Number of Animals	Date of Birth	Weight		Value	
		Initial	Final	Initial	Final
1...					
2...					
3...					
4...					
5...					
6...					
7...					
8...					
9...					
10..					

Table II—Gain in Weight

Fill in names, etc., in same order as above

Name and Registration Number of Animals	Weight												
					Gain per Month								
	Initial	1	2	3	4	5	6	7	8	9	10	Final
1...												
2...												
3...												
4...												
5...												
6...												
7...												
8...												
9...												
10..												

14.3 The record book Marietta filled out for her project encouraged her to keep track of what she fed her lambs, how much it cost, and how much weight they gained. (Waddell and Potter, "Record Book for 4-H Boys' and Girls' Livestock Clubs.")

FEEDS FED
Table III—Concentrates

Date	Kind of Feed	Amount	Price	Value
	(Enter feeds in this manner) Corn (Shelled)	60 lbs.	.01	.60
	Total			

14.3 (continued)

to a recommended feeding regimen developed at the agricultural college and provided to her by Thurber. Her 4-H bulletins and project guidelines explained with precision and order that lamb growth would follow from twice a day feeding for seven weeks, slow increases in grain rations each week, and shifting grain mixtures from oats, to oats and wheat, to oats and wheat and barley. It's not just what you eat but how you eat it. Likewise, not just what you feed but how to feed it. That is how we learn.[15]

Imagine Marietta posting the feeding schedule in the barn, right where she measured out the feed each day. Reading through the record book and bulletins she used, I could see her crossing off each week's ration as she went, leading her lambs to the scale at the end of every week and noting their weights in her record book. By the end of three months, keeping track of the lambs' feed, weights, and costs would have become a habit for Marietta.

This habit was helping her understand what the USDA called a "balanced ration." While today this term brings to mind a balanced diet, a Nutrition Facts label, or the Food Pyramid or MyPlate, to early twentieth-century agricultural specialists it meant something more specific than just a judicious mix of nutrients. For a ration to be "balanced" it needed to be both nutritionally appropriate and economically sound: it would produce the best results for the least cost. The record book revealed how balanced a ration was.[16]

Marietta had already begun to learn this lesson in the year prior to the lamb project when she had fed different parts of the family flock according to their age, sex, and purpose. Before spring lambing season she had given the pregnant ewes a special diet to prepare them for bearing healthy lambs. When each lamb was born it began to suckle, and since ewe's milk was its only food for two weeks, its mother's nutrition continued to determine its own. By ensuring the ewes had the proper diet before lambing and while the lambs were suckling, Marietta absorbed lessons about the care and feeding of mothers and babies—lessons the Extension Service hoped she would remember when she raised her own children. First the sheep, then the people.[17]

Marietta internalized the lessons of system and utility built into the balanced ration while fattening her lambs. On her way to becoming the modern farmer that embodied those same characteristics—system, utility,

14.4 County agent D. P. Thurber took this photograph of Marietta and her lamb on a visit to the Hamilton farm in the fall of 1932. Even when documenting other activities, Thurber emphasized feeding. (D. P. Thurber, "Annual Report of the Extension Agent for the Sun River Irrigation Project" [unpublished report, 1932], 70.)

calculation—she was becoming a modern eater. Because the lesson here was not just that lambs could grow into better sheep with the proper ration, but that people, too, could be transformed by food. In the 1920s, 4-H had kicked off a national health contest, based on a set of health records and scorecards for children drawn up by USDA home economists. These were explicitly modeled on the scorecards and record books that had become important and familiar instruments in judging stock, vegetables, ears of corn, and other farm products at the fair—the very same kinds of records Marietta was keeping on her lambs. So as she raised her lambs, Marietta not only monitored their food and growth—she also learned that it made sense to keep track of her own health in the same way.[18]

One warm day as the fall deepened, Thurber came out to the farm to see how Marietta was getting along. He found her kneeling in the sheep pen, trimming a lamb called Johnnie, her tanned arms holding him in place. Thurber snapped a photograph to include in his report at the end of the season (figure 14.4). His photo caption briefly explained what she was doing before describing in detail what Marietta's lambs were eating each day. He looked at the lamb and the girl, but he saw the food:

Marietta Hamilton, the only girl in the lamb fattening 4-H club is teaching Johnnie, her pet lamb to stand correctly and is blocking him out, trimming the wool on his back, head, and sides, so as to give him the appearance of being more square in conformation, and more low-set. Marietta is feeding 25 lambs, white faces, and is doing an excellent job. They are eating ½ pound per day of oats, barley and wheat mixed. They have a mixture of 30 pounds of Mono-Calcium Phosphate and 70 pounds of salt before them at all times. Their water trough is always full and clean. This pen of lambs will be "fat" in 90 days.[19]

January's fat lamb show provided a specific target for Marietta's work in the fall of 1932. She prepared Johnnie and his most promising brethren for the ring, and prepared herself. She studied up on her bulletins to prepare for the judges' questions, kept herself clean, neat, and healthy—eating well and keeping herself as nicely groomed as her lambs—and selected a good outfit for the show. All the while she weighed the lambs, Johnnie included, to see if they had reached 90 pounds, and marked the figures in her record book. When their bones were well covered in flesh and fat, and winter had come to the Fairfield Bench, Johnnie and his brethren had eaten their fill—they were ready to become food.[20]

When January arrived, Marietta and her parents loaded the lambs up for the thirty-five-mile trip from Fairfield to Great Falls. The journey took Marietta across a changing agricultural landscape. Though the yellow fields were frosted by blowing snow, the countryside still bore evidence of irrigation's fresh arrival. These were the lands that had nourished her flock, that made Johnnie and his companions Montana lamb. They made her and her project a model for the nation. At the show, both lamb and child were on display.

The Sun River club members did well in Great Falls. One of Marietta's lambs placed first in its breed class and came in third in the open class, selling for 17 cents a pound. One of her neighbor's ewe lambs weighed in at 110 pounds and won the grand champion title for the entire show (the Rainbow Hotel of Great Falls purchased her for 22 cents a pound). And the S&B Grocery paid a premium over the market price for one of the winning lambs, making it the centerpiece of the cornucopic window display that had started me on this research journey. Marietta returned home without her best lambs but with money in her pocket and new ideas in her head.[21]

I find myself unusually connected to the legacy wrought from the changes Marietta witnessed and helped create. There are a few reasons for this, one of which I'll return to in a moment. But we can look first to Marietta and the legacy her 4-H experience had for her. She didn't do the lamb show the next year. Depressed farm prices made it difficult for families—even the relatively prosperous ones of Sun River—to get started in new lines like lamb raising. She did stick with club work, though. Marietta enrolled in the dairy calf project in 1933, where her 4-H success continued. She won first place for her Guernsey in the local competition that summer. This time, filling out her 4-H livestock record was a matter of course. Her 4-H dairy bulletins at her side, pencil in hand, Marietta could easily keep track of her calf's feed, its growth, and, later on, its milk production. With her records in hand, she could make judgments about the most economical blend of feeds for the most growth or milk—a balanced ration.[22]

Marietta contributed to a new understanding of rationalized lamb-raising that created rational eating. For her rural upbringing, the lessons from raising her lambs preceded and gave meaning to the lessons for her own eating, for the way her parents raised her. She learned about a new form of eating, what we now think of as *modern*, not from the kitchen table concepts of nutrition brought to the twentieth century through consumer metrics of food purchases and the quantified labels and cans that explained what was healthy and what nutrients were present. Hers came from the experience of regimented, utility-based ways to raise animals. Hers was barnyard nutrition that made its way to the kitchen table. And, I would argue, our kitchen table nutrition today still has a whiff of the barnyard about it.

As Marietta was aging out of 4-H, the children who followed were getting the food and health message more explicitly across all their projects. Whether they were raising vegetables for the family garden, cooking meals or sewing clothes, raising fields of grain or herds of livestock, 4-H'ers were learning to think about food in a modern way, as a set of nutrients with different properties and effects. Like Marietta, they had absorbed this lesson by first caring for plants and animals. The analogy instilled from record keeping was clear: just like her lambs, Marietta's health depended on what she ate and how. By raising a lamb, she learned a systematic approach to food for both her livestock and herself.[23]

But there is a coda to this story, one that illustrates the limits of modern food. What you eat has the power to change you, but you don't always have the power to change what you eat. And the measure of what you eat depends entirely upon the state of nutritional knowledge at any given point in time. Today's balanced ration may be tomorrow's deficient diet.

Marietta would soon apply these lessons in her own family life. At age seventeen she got married to Maynard Lehman, ten years her senior, in a small ceremony in Missoula; by 1938 she was up in the Bitterroots, starting a family of her own.[24] The next June she gave birth to a son, Gary Edwin. You can imagine her recalling the importance of good nutrition for pregnant ewes from her lamb-raising bulletins and trying to eat as well as she could manage. I could see her breastfeeding this child, thinking back on what she had read about the importance of colostrum for her newborn lambs. But one night in September tragedy struck, and her three-month-old baby died in his sleep. The doctor said it was thyroid trouble; the death certificate read *status lymphaticus*—essentially, an unexplained sudden death. Little Gary Edwin was buried in a cemetery in Ravalli, in the Bitterroot Valley, where Marietta and Maynard—and, eventually Marietta's parents too—had settled, nearly 200 miles from Sun River.[25]

I found Marietta's grave during my research for the larger story of which Marietta's lamb is a part. She was buried beside her husband and infant son. I tried to imagine how this young woman, just barely eighteen, might have understood this tragic event when her son died. Would she have worried that she didn't eat well enough to make a healthy baby, or that she hadn't fed him properly in those early months? If she and Maynard were struggling to make ends meet—not out of the question in 1938—might she have been faced with impossible choices about how to nourish herself and her family? Did her 4-H training in food economy and rationing help her do the best with what she had, or was it useless in the face of financial hardship and personal tragedy?

These sorts of questions had become acutely personal for me. I had just learned I was pregnant and was thinking about food and nutrition all the time. My doctor gave me a big box of prenatal vitamins. My mother sent me bottles of supplements to support fetal brain development and, on every phone call, asked if I was getting enough protein in my diet. My eating responsibilities had just doubled, and suddenly everyone was interested. Like the farm women and girls of the early twentieth century,

I had become a key target of nutritional education. What I cooked and ate mattered in a new way.

Marietta's experience as a mother and a lamb raiser affected her deeply. The remainder of her life offered clues about her thoughts and feelings that suggest the endurance of these early lessons and experiences. She and Maynard never had any more children. They eventually left rural Montana for Spokane—one of the railroad centers that gathered the produce of the inland West, from timber to wheat to the forage crops the Hamiltons had grown—and later Denver. Though they moved to the city, their lives on the western frontier continued to shape them as they aged. In the 1980s they began writing Western novels together under the name M & M Lehman, most of which were published as audio books through a Spokane recording studio. I've listened to some excerpts, and while they're not the height of literature, they make it clear that the struggles—financial, agricultural, personal—of Marietta's youth and early adulthood had a lasting impact. Several of the books—with titles like *Sheep Country* and *Wool on the Drift Fence*—feature a plucky heroine who has to save the ranch after the death of a family member. Though they hew to the Western genre, they have a spark of personal experience about them that defies the formula. The girl posing with her lamb in Great Falls in 1933 is still there, looking calmly into the camera.[26]

NOTES

1. D. P. Thurber, "Annual Report of the Extension Agent for the Sun River Irrigation Project" (unpublished report, 1933), 39, Montana State University Extension Service Records, box 72, folder 21, Accession 00021, Merrill G. Burlingame Special Collections, Montana State University Library, Bozeman (hereafter Burlingame Special Collections).

2. Thurber, "Annual Report" (1933), 38.

3. Thurber, "Annual Report" (1933), 43.

4. In 1932 there were 925,612 4-H Club members in the United States; see Florence L. Hall, "4-H Club Work, 1932," USDA Extension Service Circular 192 (September 1933), 1.

5. Robert Autobee, "Sun River Project," U.S. Bureau of Reclamation, 1995, https://www.usbr.gov/projects/pdf.php?id=198; D. P. Thurber, "Annual Report of the Extension Agent for the Sun River Irrigation Project" (unpublished report, 1930), Montana State University Extension Service Records, box 72, folder 18, Accession 00021, Burlingame

Special Collections; Thurber, "Annual Report of the Extension Agent for the Sun River Irrigation Project" (unpublished report, 1931), Montana State University Extension Service Records, box 72, folder 19, Accession 00021, Burlingame Special Collections; Thurber, "Annual Report of the Extension Agent for the Sun River Irrigation Project" (unpublished report, 1932), 65, Montana State University Extension Service Records, box 72, folder 20, Accession 00021, Burlingame Special Collections; Thurber, "Annual Report" (1933), 36–37; "Sun River Farms to Be Opened in June for Homesteading," *The Independent Record* (Helena, MT), May 23, 1931, 6; "Ex-Service Men Ask for Every Unit of Greenfield Project," *The Independent Record* (Helena, MT), October 1, 1931, 7; "How to Get a Good, Irrigated, Diversified Farm Home," *The Montana Standard* (Butte, MT), February 22, 1931, 22; 1930 United States Federal Census for Kevin, Toole County, Montana, Roll 1263, 1A, enumeration district 0009, FHL microfilm 2340998, accessed via ancestry.com, January 2018. This scene is also based on contemporaneous USDA films and photographs depicting the work of extension agents and my reading of extension agents' reports from around the country, including collections at the National Archives, the Cornell University Archives, the University of Wisconsin Archives, the Mississippi State University Archives, and the Montana State University Archives.

6. Thurber, "Annual Report" (1933), 36–37; Thurber, "Annual Report" (1932), 56; D. E. Richards, "The 4-H Club Beef Calf," Montana Extension Service Bulletin 118, January 1932, 2; John Dexter, ed., "An Agricultural Program for Montana," Montana Extension Service Bulletin 84, May 1927, 13, 14, 49; Sarah T. Phillips, *This Land, This Nation: Conservation, Rural America, and the New Deal* (New York: Cambridge University Press, 2007); Autobee, "Sun River Project," 24–29; R. B. Tootell, "An Inventory of Montana Irrigation Projects," Montana Extension Service Bulletin 124, September 1932, 53–55.

7. Thurber, "Annual Report" (1933); J. M. Cooper, "Sheep of the Columbia Type Well Adapted to Intermountain Region," in *Yearbook of Agriculture, 1928* (Washington, DC: Government Printing Office, 1929), 540–541.

8. Franklin M. Reck, *The 4-H Story: A History of 4-H Club Work* (Ames: Iowa State College Press, 1951); Thomas Wessel and Marilyn Wessel, *4-H: An American Idea, 1900–1980* (Chevy Chase, MD: National 4-H Council, 1982); Gabriel Rosenberg, *The 4-H Harvest: Sexuality and the State in Rural America* (Philadelphia: University of Pennsylvania Press, 2016); David B. Danbom, *The Resisted Revolution: Urban America and the Industrialization of Agriculture, 1900–1930* (Ames: Iowa State University Press, 1979); Laura Lovett, *Conceiving the Future: Pronatalism, Reproduction, and the Family in the United States, 1890–1938* (Chapel Hill: University of North Carolina Press, 2007); Sarah Stage and Virginia B. Vincenti, eds., *Rethinking Home Economics* (Ithaca, NY: Cornell University Press, 1997), esp. chap. 7; Kathleen R. Babbitt, "Legitimizing Nutrition Education: The Impact of the Great Depression," in Stage and Vincenti, *Rethinking Home Economics*, 145–162.

9. Thurber, "Annual Report" (1930), 27.

10. This is based on research in the extension collections already cited, as well as Stage and Vincenti, *Rethinking Home Economics*.

11. These features of the consumer world have been well canvassed by historians. See, for example, Harvey A. Levenstein, *Revolution at the Table: The Transformation of the American Diet* (New York: Oxford University Press, 1988); Helen Zoe Veit, *Modern Food, Moral Food: Self-Control, Science, and the Rise of Modern American Eating in the Early Twentieth Century* (Chapel Hill: University of North Carolina Press, 2013); Rima D. Apple, *Vitamania: Vitamins in American Culture* (New Brunswick, NJ: Rutgers University Press, 1996); Nick Cullather, "The Foreign Policy of the Calorie," *American Historical Review* 112, no. 2 (April 2007): 337–364.

12. Miriam Birdseye, "Growth Work with 4-H Clubs," USDA Extension Service Circular 14, October 1926; Ralph A. Felton and Nina V. Short, "Rural Health," Cornell Extension Bulletin 187, November 1929; *Food for Reflection* (Washington, DC: USDA Extension Service, 1921), 33-146, Motion Picture Films, ca. 1915–ca. 1959, Records of the Extension Service, Record Group 33, National Archives and Records Administration, College Park, MD.

13. R. L. Waddell and Charles E. Potter, "Record Book for 4-H Boys' and Girls' Livestock Clubs," Montana Extension Service Bulletin 76, March 1926; D. E. Richards and Louis Vinke, "Feeding Lambs in Montana," Montana Extension Service Circular 26, July 1931; "Sheep Club Rules and Regulations," Montana Extension Service in Agriculture and Home Economics, no. 34, March 1919; "Daily Record Book of Boys' and Girls' Club Projects," Montana Extension Service in Agriculture and Home Economics, no. 23, May 1917; D. E. Richards, "The 4-H Club Beef Calf," Montana Extension Service Bulletin 118, January 1932; J. O. Tretsven, "4-H Club Dairy Calves," Montana Extension Service Bulletin 126, October 1932; I. M. C. Anderson, "4-H Sheep Club Manual," Montana Extension Service Bulletin 134, May 1933; Pauline Bunting, "Health: The Fourth H," Montana Extension Service Bulletin 144, April 1935; Pauline Bunting, "The Health H," Montana Extension Service Bulletin 154, October 1937.

14. Anderson, "4-H Sheep Club Manual," 15; Thurber, "Annual Report" (1933), 37; *Bob Farnum's Ton Litter*, dir. Fred W. Perkins and James R. Wiley (USDA Educational Film Service, 1923), 33-232, Motion Picture Films, ca. 1915–ca. 1959, Records of the Extension Service, Record Group 33, National Archives and Records Administration, College Park, MD; Keira Butler, *Raise: What 4-H Teaches Seven Million Kids and How Its Lessons Could Change Food and Farming Forever* (Berkeley: University of California Press, 2014).

15. Waddell and Potter, "Record Book for 4-H Boys' and Girls' Livestock Clubs"; Thurber, "Annual Report" (1933), 25, 36–37.

16. A. L. Knisely, "Balanced Rations for Stock," *Cornell Reading-Course for Farmers*, series 2, no. 6 (November 1901), 94–95.

17. Anderson, "4-H Sheep Club Manual," 5, 11; Richards and Vinke, "Feeding Lambs in Montana"; Helen Monsch, "Feeding Babies and Mothers of Babies," Cornell Bulletin for Homemakers, Bulletin 300, June 1934; Martha Mae Hunter, "Diet for Mother and Child," Montana Extension Service Bulletin 67, June 1923; "The Care of the Baby," Montana Extension Service Bulletin 18, April 1917; Ralph A. Felton and Nina V. Short, "Rural Health," Cornell Extension Bulletin 187, November 1929, 8.

18. Richards and Vinke, "Feeding Lambs in Montana," 13; Waddell and Potter, "Record Book for 4-H Boys' and Girls' Livestock Clubs"; Reck, *The 4-H Story*, 182; Helen Parsons, "Score Cards for Children," Wisconsin Extension Service Special Circular, June 1924; Bunting, "Health."

19. Thurber, "Annual Report" (1932), 70.

20. Richards and Vinke, "Feeding Lambs in Montana," 2, 11; Waddell and Potter, "Record Book for 4-H Boys' and Girls' Livestock Clubs," 4, 11.

21. Anderson, "4-H Sheep Club Manual," 19; "Woolmen Hope to Enjoy Better Era," *The Independent Record* (Helena, MT), January 19, 1933, 1, 10; "Grand Champion 4-H Lamb Chosen," *The Montana Standard* (Butte, MT), January 19, 1933, 11; Thurber, "Annual Report" (1933), 36–37.

22. D. P. Thurber, "Annual Report of the Agricultural Extension Agent for the Sun River Irrigation Project" (unpublished report, 1934), 71–73, Montana State University Extension Service Records, box 72, folder 22, Accession 00021, Burlingame Special Collections; Thurber, "Annual Report" (1933), 35; Tretsven, "4-H Club Dairy Calves."

23. Reck, *The 4-H Story*, 182; Waddell and Potter, "Record Book for 4-H Boys' and Girls' Livestock Clubs"; Bunting, "Health"; Bunting, "The Health H."

24. Marriage license for Maynard Lehman, 1938, in *Montana, County Marriages, 1865–1950* (Salt Lake City, UT: FamilySearch, 2013), 104, no. 1264; U.S. Bureau of the Census, *Sixteenth Census of the United States, 1940* (Washington, DC: National Archives and Records Administration, 1940), roll T627_2227, 6B, enumeration district 41-15, accessed via ancestry.com, September 20, 2017.

25. Death certificate for Gary Edwin Lehman, September 20, 1939, certificate no. 2816, Montana County Births and Deaths, 1830–2011, Montana State Historical Society, Helena, MT, accessed via ancestry.com, September 20, 2017; Population schedule for Stevens, Ravalli, Montana, roll T627_2227, 6B, enumeration district 41-15, in U.S. Bureau of the Census, *Sixteenth Census of the United States, 1940*, T627, accessed via ancestry.com, September 20, 2017; population schedule for Hot Springs, Sanders, Montana, roll T627_2229, 2B, enumeration district 45-17, in U.S. Bureau of the Census, *Sixteenth Census of the United States, 1940*, T627, accessed via ancestry.com, September 20, 2017; Ann Dally, "Status Lymphaticus: Sudden Death in Children from 'Visitation of God' to Cot Death," *Medical History* 41, no. 1 (January 1997): 70–85.

26. *U.S. Public Records Index, 1950–1993*, vol. 1 (Provo, UT: Ancestry.com Operations, 2010); *U.S. Social Security Death Index, 1935–2014* (Provo, UT: Ancestry.com Operations, 2011); "Marietta Fern Hamilton Lehman (1920–2003)—Find a Grave Memorial," accessed November 16, 2018, https://www.findagrave.com/memorial/84724389 /marietta-fern-lehman; "Maynard Milton Lehman (1910–2013)—Find a Grave Memorial," accessed November 8, 2018, https://www.findagrave.com/memorial/120482062 /maynard-milton-lehman; "Books in Motion Search—Lehman," accessed November 8, 2018, https://www.booksinmotion.com/index.php?route=product/search&search =lehman.

EPILOGUE

In the story of modern food, the 1940s is not a beginning but a middle.

At that middle, then, the food system looked like this, the product of the last half-century of change:

Time and space had shrunk, to the benefit of some and the ruin of many. The Yokuts people and the thriving California lake they lived on faced existential threats from financial manipulation and from wheat bound for England. In England, that bread took on a color that denoted class connections. Filipinos lived under the uncanny rule of Americans dedicated to making food from the land while pretending that the people who lived there had never made anything worth eating. Competition for European beer markets had led to innovation in the form of a global golden pilsner. Stevedores wrestled with heavy crates of rotting fruit while merchants outran decay with their pushcarts.

Regulators and businesspeople fought each other for public trust. Corn syrup had become sugar, depending on who you asked. Standards governed food and led to a search for standardized consumers as boxed breakfast cereal and other processed foods spread. Black freedom fighters opened grocery stores in pursuit of justice. Black performers gained influence and found their images appropriated to sell fruit. Long-trusted opponents of canned food now used their celebrity to give the industry legitimacy.

And scientific categories, or changes called science, underlay all. New tools and measures gauged purity and whiteness in sugar and in people. People who wanted to reduce foods down to their component parts created predigested meats. People who found meat abominable tried their best to imitate it. Children who lived on farms learned to fatten, measure, and enter into competition the animals that would become standardized meat, as the children themselves learned about scientized nutrition.

All this led to a different food system. At the heart of it, for better and for worse, were global interconnections, widespread human and environmental damage, industrial production, scientific development, unequal access, and narrow ideas about health.

Food has changed since the 1870s. But it has changed in ways that follow the development of modern food around the turn of the twentieth century, the fruits of which are still with us. Writers about food have been warning for a generation or more that our food system was unsustainable, on the verge of calamity. We no longer need to make predictions. We are there.

It didn't take a global pandemic to create problems for food, farming, and distribution, but the experiences from 2020 and after certainly laid the problems bare. We're writing this in the summer of 2020, unsure where things go from here, but clearer eyed about how we got here. When infectious disease shut down the United States, in particular, the various pieces of our food system collapsed into themselves. Here was a disease born in part of modernity, shaped by human–animal relationships, and spun around the globe amid fractured political wills to address its devastation. And then the attempted solutions to the havoc wrought on the food system have been rooted in more modernity. It is the age of doubling down.

Time and space have collapsed to ostensibly feed the world. While factory and transit workers could not safely work, those who could remain separate drew veils around physical distance. Mills and restaurants and wholesalers sold directly to consumers because the internet—that forlorn vehicle for the current annihilation of space and time—meant a physical retail location wasn't necessary to sell. Some people risked their lives to deliver food ordered on apps to others who understood their duty to society as staying inside their homes.

Faces and human hands once again became as worrying as they had been in the early 1900s, when factory production and machine operation offered the solution to a fear of newfound germs. In our time we covered our faces with masks. Digital technology as intermediary meant that those ordering foods could ask for touchless deliveries, an option not available to cooks or delivery people but which, just maybe, protected them too. Technology as luxury meant that the illusion of human removal from the process of food production could be maintained for some while disease attacked poor communities of color.

For the wealthy, cooking trended. It took on an allure but primarily as a way to pass abundant and luxurious leisure time that could no longer be devoted to in-person social gatherings. Sourdough bread became famous. Despite a small resurgence of gardening that mirrored prior efforts in times of crisis, when it came to trust, Americans as they had before turned to packaged food. Processed food sales increased by more by a third after the pandemic started, offering comfort and safety to nervous buyers. We doubled down for lack of a better route forward.[1]

And because food was the stuff of capital markets and international trade, as had been put into place between the 1870s and 1930s, the people that those systems hurt also suffered just in trying to provide and purvey food. The pandemic hit working-class people and Black, Indigenous, and other people of color harder than it hit wealthy white people. It hurt all restaurants, reserving especial zealousness for those owned by Black women. Black people have worked together to try and avert the worst, starting programs to keep Black-owned restaurants in business during the crisis. Even Black farmers found themselves disproportionately excluded from urban farmers' markets.[2]

Much of the change in food structures that the pandemic brought on is bad for many people. Some of it is good, depending on your point of view, a testament to human ingenuity that has fed millions. All of it is modern. We see a continued trend toward making individual homes the clearinghouse of culinary commerce for those who can afford it, toward validating trust in brand names and the invisibility of human laborers, toward equating safety with science. All of this reflects the foundations that were laid in the period described in the chapters you've just read. By the 1930s, the trends that have only intensified in the face of our twenty-first-century global pandemic were in place, ready to shape the next century of American eaters.

If, as Tom Finger writes in our opening essay, ghosts haunted the creation of modern food, they have not been exorcised. They continue to lurk amid efforts to rethink where our food comes from and what we do with it. This is not surprising, because we are still living the same story of modern food. The question now is if, whether, and how that story ends— and what comes next.

NOTES

1. Michael Moss, "Has Pandemic Snacking Lured Us Back to Big Food and Bad Habits?," *New York Times*, June 16, 2020, sec. Well, https://www.nytimes.com/2020 /06/16/well/eat/pandemic-snacking-junk-food-habits-eating-weight.html.

2. Evan Garcia, "'Black People Eats' Showcases Black-Owned Restaurants in Chicago and Beyond," WTTW News, June 29, 2020, https://news.wttw.com/2020/06 /29/black-people-eats-showcases-black-owned-restaurants-chicago-and-beyond; Jenny Splitter, "Largest D.C. Farmers Market Repeatedly Denied Spots to Black Vendors, Farmers Allege," Forbes.com, June 15, 2020, https://www.forbes.com/sites /jennysplitter/2020/06/14/racism-dc-largest-farmers-market/#5d9164ee575b.

ACKNOWLEDGMENTS

Books and stories have beginnings, middles, and ends. So does the writing process. Here at the end of this collection, we can acknowledge how we got here. To paraphrase E. M. Forster, the basic story is that the contributors signed on and then the book came together; the plot is that the contributors signed on and then the book came together because of a common interest in writing beyond the academy.

Back at the beginning of this project, we were encouraged by so many writers and historians who pushed us to think more deliberately about the writing craft. That encouragement got us to the middle of the writing process, when we gathered as a group to talk about writing style, voice, prose, and audience. We want to acknowledge the range of people who helped us put the pieces of this book in place at that point. In particular, the writer, editor, and teacher Helen Rubinstein helped us think through two key questions: Who are we writing for? How do we write for them? With Helen's encouragement, we also shared the writing of those who have inspired us to produce intellectually grounded, evocatively written work—people including Joan Didion, Roxane Gay, James Goodman, Jill Lepore, John McPhee, Susan Orleans, Rebecca Solnit, and, yes, the list goes on. With Helen's guidance, we sought to rethink how we cued our points, how we cast our voices, and how we could find ways to share the most interesting parts of our historical work in pieces relatively briefer than books and articles. At a retreat for the writers who signed on to the collection early, we had help from Jeremy Zallen and Rochelle Greenidge, as well as our gracious hosts at the Christmas City house, Jay and Mary Ellen. Rachel Hogan Carr and the staff of the Nurture Nature Center in downtown Easton, Pennsylvania, provided a forum for our night of spoken-word storytelling during the weekend retreat. And even before all of that, funding from Lafayette College proved crucial. We thank Kristen Sanford, Serena Ashmore, and the Engineering Studies Program; Josh Sanborn and the History Department; Abu Rizvi and the Office of the Provost; Dave Brandes and the Programs in Environmental Science and Environmental Studies; the Class of '74 Endowment for the Forum on Technology and the Liberal Arts; and Scott Hummel and the Engineering Division, all at Lafayette. We're grateful for your help.

Conversations with others in the middle of our process and since pointed us in useful directions too. John Warner, Aaron Sachs, Rebekah Pite, and James McWilliams helped us think through history and writing, and in some cases looked at drafts. Heather Lee's early involvement at our retreat helped to improve first versions of several chapters. We also thank Beth Clevenger, Robert Gottlieb, Nevin Cohen, and Anthony Zannino at the MIT Press for taking the project on board and three anonymous reviewers for helping encourage its final shape.

The beginnings now feel distant, the middle spanned the period before and now through the pandemic, and the end of the story is the book in your hand. We edited it as true coequals, sharing planning, preparation, and publication with constant congeniality and good will. And in the end, we thank our families and friends for their support along the way. Thank you to Chris, Harper, and Alex in Easton, Pennsylvania; Madeleine and Enid in Princeton, New Jersey; and Justin, Nancy, and Mira in Blacksburg, Virginia. Our stories would have no meaning without them.

CONTRIBUTORS

Benjamin R. Cohen is an associate professor at Lafayette College. He is the author most recently of *Pure Adulteration: Cheating on Nature in the Age of Manufactured Food* (2019).

Thomas D. Finger teaches American and environmental history at Northern Arizona University. His research focuses on environmental histories of food, water, and energy systems. In his teaching and research, Finger looks beyond national borders to ask how larger communities of people, plants, and animals live within human production chains.

David Fouser is a visiting instructional assistant professor at Chapman University and an adjunct faculty member at Santa Monica College and Laguna College of Art & Design. He completed his PhD in 2016 at the University of California, Irvine, on the cultural and environmental history of wheat, flour, and bread in Britain and the British Empire in the nineteenth century. He currently balances teaching and continuing his research on the globalization of Britain's bread.

Lisa Haushofer is a senior research associate at the Institute of Biomedical Ethics and History of Medicine at the University of Zurich. She holds a PhD from the Department of History of Science at Harvard University and an MD from the University of Witten-Herdecke, Germany. Her research examines the intersection of the histories of science, food, and economic life. She is currently writing a history of "wonder foods." She is a member of the editorial collective of *Gastronomica: The Journal for Food Studies* and a review editor at *Social History of Alcohol and Drugs*.

Michael S. Kideckel completed his PhD at Columbia University in 2018. His book project, "Fresh from the Factory: Breakfast Cereal, Natural Food, and the Marketing of Reform, 1890–1920," is currently under contract with Oxford University Press. He teaches history at Princeton Day School in New Jersey.

Faron Levesque is a food justice activist and historian currently based in their hometown of Memphis, Tennessee, where they run a community teaching kitchen for AOVS Urban Farm. Faron holds an MA in History from the University of Wisconsin-Madison and a certificate in the Public Humanities. Their research, teaching, and writing interests include the history of food, political education, and the Queer South.

William Thomas Okie is associate professor at Kennesaw State University and associate editor of the journal *Agricultural History*. He is the author of *The Georgia Peach: Culture, Agriculture and Environment in the American South* (2016).

René Alexander D. Orquiza Jr. is an assistant professor of history at Providence College. He is the author most recently of *Taste of Control: Food and the Filipino Colonial Mentality under American Rule* (2020).

Jeffrey M. Pilcher is a professor of history and food studies at the University of Toronto. His books include *Planet Taco: A Global History of Mexican Food* (2012) and *Food in World History*, 2nd ed. (2017). He is an editor of the journal *Global Food History*.

Adam Shprintzen is an associate professor of history and honors director at Marywood University. He is the author of *The Vegetarian Crusade: The Rise of an American Reform Movement, 1817–1921* (2013).

David Singerman is an assistant professor of history and American studies at the University of Virginia. He is writing a book about science, corruption, and monopoly in the American sugar empire at the turn of the twentieth century.

Tashima Thomas is a visiting assistant professor at Pratt Institute. She is the author of *Edible Extravagance: The Visual Art of Consumption in the Black Atlantic* (2021).

Amrys O. Williams is the executive director of the Connecticut League of History Organizations. Previously, she was the oral historian and associate director of the Center for the History of Business, Technology, and Society at the Hagley Museum and Library. She also taught history at Wesleyan University, where she was faculty advisor to the student farm.

Anna Zeide is an associate professor of history and director of the food studies program at Virginia Tech. Her first book, *Canned: The Rise and Fall of Consumer Confidence in the American Food Industry* (2018), won a 2019 James Beard Award.

INDEX

Food, Health, and the Environment

Series Editor: Robert Gottlieb, Henry R. Luce Professor of Urban and Environmental Policy, Occidental College